HAPPOLD

THE CONFIDENCE TO BUILD

Professor Derek Walker
Dr Bill Addis

Acknowledgements

This monograph seeks to explore the life and work of Sir Edmund (Ted) Happold and the legacy he has left to his friends and colleagues at Buro Happold. In just twenty one years the fellowship of Buro Happold has become one of the country's pre-eminent engineering practices, with an international reputation for excellence and an enviable portfolio of projects which demonstrate their skills.

The authors are deeply indebted to Eve Happold and her sons Matthew and Tom for their active help and sensitive support throughout the preparation of this book from inception to completion. Their solidarity as a family unit was as inspiring as their love for Ted himself. We hope we have captured the spirit of the man, his restless energy, his humanity and his pre-eminence as one of the centuries most distinguished engineers.

We would also wish to thank many of his partners, colleagues and friends who have given generously of their time, adding richness, colour and humour to our appreciation of the man. At Buro Happold, Michael Dickson, Ian Liddell, Terry Ealey, Rod Macdonald, John Morrison, Rodger Webster, Mike Cook and Peter Moseley. At Arup's, Povl Ahm, Richard Haryott, Poul Beckmann, Martin Manning, Derek Sugden, Michael Bussell, Malcolm Millais and Turlogh O'Brien. In Germany, Frei and Ingrid Otto, Rolf Gutbrod and Christoph Ingenhoven. In the United States, Bob and Roberta Silman. In Norway, Michael Barclay. In Britain, Richard Rogers, Trevor Dannatt, Norman Foster, Ted Cullinan, John Young, Spencer de Grey, Graham Watts, Richard Burton, Robin Nicholson, Ken White, Ted Hollamby, Jack Bonnington, Max Fordham, Kit Allsopp, Stephen Ledbetter, Michael Barnes, Bryan Harris, Hugh Johnston and Christopher Frayling.

Barbara Towers, Carole Light, Helen Elias and Marilynn Osment of Buro Happold have provided an avalanche of material from the Buro Happold archives; The Arup Journals offered excellent technical coverage of projects involving Ted Happold. Pauline Shirley, Photographic Librarian, and her colleagues from the Arup's photographic library, and Frei and Ingrid Otto from the Atelier Warmbronn archives, have generously supplied us with a wealth of material for the sections on Work with Arup's, Islam - A Journey, and Lightweight Structures.

We have also had the use of the Richard Rogers archive, Derek Walker Associates archive, Foster Associates archive, Trevor Dannatt archive, Ted Hollamby archive, Renton Howard Wood Levin archive, and the Happold family archive; who all provided photographic material for the book.

We would also like to thank Brian Tattersfield, a great admirer of Ted's chutzpah, who designed the cover, Michael Ridden for preparing line drawings for a number of the projects, Teresa Martin for typing and revising all the written text, Annie Bridges our copy editor, and Dennis Crompton who designed the Ted Happold exhibition 'On Being An Engineer' which provided material for a 'Patterns' publication and for this book.

Though a small number of photographers cannot be traced, we would like to credit the following photographers whose work has been illustrated, specifically from the major archives mentioned above. They include Marc Ribaud, Henk Snoek, Richard Einzig, Richard Davies, John Donat, Ken Kirkwood, John and Jo Peck, Ian Lambot, Trevor Dannatt, Jeremy Hall, Frei Otto, Michael Dickson, Samantha Larrance, Ian Liddell, Terry Ealey, Ted Hollamby, Jan Walker, Bodo Rasch, Theo Crosby, Derek Walker, Bill Addis, Martin Charles, Malcolm Tucker, Harry Sowden, Peter Cook, Peter Webb, Mandy Reynolds and Crispin Boyle.

Design and layout by Derek Walker and Samantha Larrance.
Cover design by Brian Tattersfield.
First published in Great Britain in 1997 by Happold Trust Publications Limited.
ISBN No 0-419-24070-5

Contents

Printed and bound by P J Reproductions, London, England.

Sir Edmund Happold 1930–1996

ROYAL COLLEGE OF ART

RECTOR AND VICE-PROVOST

PROFESSOR CHRISTOPHER FRAYLING

The first time I met Ted Happold was at the University of Bath in autumn 1976, when he had just been made Professor of Building Engineering and I was a lecturer in the history of ideas. The University of Bath had started life in the previous decade as a College of Advanced Technology, and there was still a strong sense of 'them and us' in the relationship between the Schools of Humanities and Engineering. Humanities people were thought by engineers to have both feet planted firmly in the air, to be interpreters rather than performers; Engineering people were thought by historians and social scientists to have both feet planted too firmly on the ground, and to get their rather humourless kicks from reading slides rules on site. Ted, it was immediately apparent, was not a man who fitted either stereotype.

He chuckled a lot. He wore his spectacles on his forehead. He brushed his hair forward, like a silver-topped Beatle. He drove a fast car. And he believed passionately that 'a world which sees art and engineering as divided is not seeing the world as a whole'. We talked in the senior common room, over apple doughnuts served by a chatty Welsh woman called Blodwyn, about the things that architects and engineers should have in common but seldom did; about engineering design as a technological idea ('at its best an art, in that it extends people's vision of what is possible ... but not in the sense of a visual style or fashion'); and about the eighteenth-century engineer John Smeaton who, like Ted, came from Leeds, and who combined his practice as an engineer with study of philosophy long into the night. We also talked about the importance of interdisciplinarity – 'student architects and engineers should share projects' – and the dangers of overdoing the distinction between C P Snow's *Two Cultures*. Here was a larger-than-life man who came from 'the science side' (he was educated as a geologist and a civil engineer) and who knew a great deal about both history and art: he spoke equally ebulliently about all of them. A bridge-builder in every sense. I subsequently discovered that he was a Quaker and a fan of the writings of William Penn, the pioneer who wrote admiringly of people who didn't 'hide their light under a bushel, but set it on a table in a candlestick'. Since by then I was combining my lectureship at Bath with a part-time appointment at the Royal College of Art – an institution which teaches postgraduates to design both tables and candlesticks – Ted's excitement again proved infectious. I still think of him every time I drive through Pennsylvania, a little village on the A46 just outside Bath, which I do at least twice a week.

The last time I met Ted Happold was in the Royal Albert Hall, at the Royal College of Art Convocation ceremony of July 1993 where he was being awarded a Senior

Beaubourg, Paris

Fellowship. I was by then the College's Public Orator – a role which traditionally involves celebrating the great achievements of honorands in a rhetorical style which includes a lot of outrageous puns. Once the assembled parents, graduates and dignitaries have grasped the fact that they are not at an especially solemn occasion, all tends to go well. Before that moment – and it usually is one particular moment – being the Public Orator in 'the nation's village hall' is a little like performing in Brighton, off season, on a wet Thursday afternoon. It was during my oration for Ted Happold that everyone began to get the message; and it was the twinkle in Ted's eye which signalled to the congregation that they were allowed to have a good time.

Professor Edmund Happold – known to his friends as Ted – as head of the School of Architecture and Building Engineering at the university of Bath, as the boss of the highly regarded Buro Happold consulting engineers, and as the man who helped to enable such ambitious buildings as the Pompidou Centre, the Conference Centre at Mecca, the Sports Hall at Jeddah, the Diplomatic Club at Riyadh, the Tsim Sha Tsui Cultural Centre and Kowloon Park in Hong Kong, the Aviary in Munich Zoo and over thirty others of their ilk to stand up and stay in one place, has spent his entire professional career since 1960 building strong bridges between the world of engineering and the world of design, and hapholding the relationship between them. Also, giving meaning to that much misused phrase 'an elegant solution'. Indeed, many of the innovations commonly attributed to some of our most famous architects should really be credited to Happold, as a recent exhibition at the Royal Academy made clear by not deigning to mention the fact. He even re-designed the graduation ceremony in the yard of Harvard University, so he is no doubt watching today's fancy-dress party with a keen professional eye. He's currently reached the dizzy heights of Master of the Faculty of Royal Designers for Industry, and vice-president of the Royal Society of Arts and in his time he's been awarded no less than nine honorary fellowships of learned institutions – but never until today, for some unaccountable reason, a Senior Fellowship. In the often uneasy marriage between engineering, architecture and design, it is clear that the words 'to hap and to hold' should be added to the ceremony forthwith – after all, he lists as his recreations 'engineering and the family' – and equally clear that Ted Happold should be warmly welcomed into our fellowship. Professor Edmund Happold.

In his reply on behalf of the honorands – delivered with his Fellowship hood slung at a jaunty angle over his shoulder, and his tie as usual at half mast – Ted spoke eloquently and genially of the importance of designers and engineers understanding each other, of the dangers of mere styling in the 'designer decade' and of the importance to him of combining professional practice, education, research and institutional affairs with a deep sense of humanity in all these things. Wherever Prince Albert was at the time, he must surely have been smiling. Four years later, there's a Ted-sized hole in the worlds of Engineering and Design: and it is a big one.

Christopher Frayling, August 1997

CONSTRUCTION

A Singular Life

> And this is the comfort of the good that the grave cannot hold them, and that they live as soon as they die, for death is no more than a turning of us over from time to eternity. Death, then being the way and condition of life, we cannot love to live if we cannot bear to die.
>
> William Penn, Quaker 1682

Indeed the grave does not hold Ted. The richness of his life, the sum of his achievements, the love and devotion of his family and friends, the recollection of a thousand memories, help to bridge a void that seemed unbridgeable when he died. He was a very singular person, loyal, true, energetic, charismatic and with a deep and sincere love of humanity. He enriched the lives of those who knew him by virtue of his generosity of spirit and the quality of his intelligence.

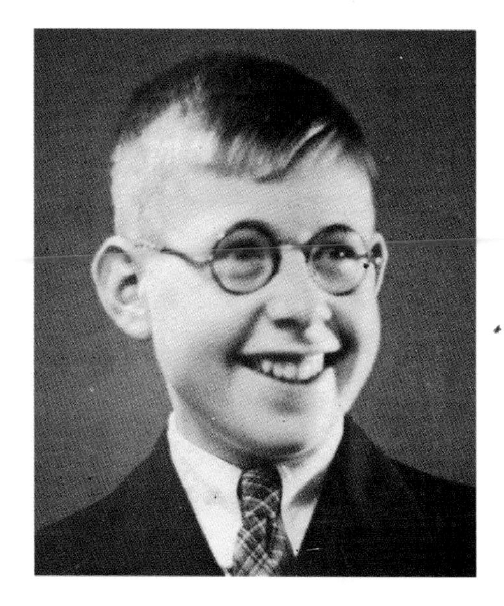

I must speak personally about him because we shared the same roots, the same time and the same values. I was proud to be his friend and to enjoy an intimacy born out of endless conversations on a myriad of topics ... It might be the family, a current project, the state of cricket, the theatre, education, D'Arcy Thompson, building physics, an obscure painter, catanaccio, building control in Hutton le Hole, Feng Shui or Alan Bennett. To all of them he brought a bubbling enthusiasm for life and all that it contained and it is this endless and insatiable curiosity, analysis and perceptive insight that I miss most.

One of Ted's last gifts to me the Christmas before he died was *City Lights*, Keith Waterhouse's autobiography of childhood days in Leeds. I read it cover to cover on Christmas Day and retraced and embraced the milieu and the warmth and wisdom of the circles we moved in as children ... The streets we shared, the buildings we revered, the magical atmosphere of the Leeds market hall, the formidable strength and solidity of the Brodrick buildings, the sanctity of Headingley and the flickering ghosts of summers past, and the endless pilgrimages to the Dales and Fountain's Abbey.

Ted's personality and sense of purpose was fashioned in the demanding climate of his domestic life in that northern city. His father and mother were a formidable pair – deeply committed to the Quaker faith and a brand of Fabian socialism ... Life was a serious crusade but the high-minded discussion around the family hearth by their ever increasing circle of friends and colleagues was lightened by Margaret Happold's gregarious and outgoing nature. She had a great sense of fun and a wide range of interests ... Indeed her need for mental stimulation, travel and social justice mirrored the characteristics which dominated Ted's own life. Ted's father, Frank Charles Happold, was less ebullient than his wife and perhaps less outgoing. He was a very distinguished academic with a full range of degrees: Masters, PhD and Doctor of Science. He became the first Professor of Biochemistry at the University of Leeds and was the Chairman of the Biochemical Society (1963–5) and in 1964 became the first Chairman of the Federation of European Biochemical Societies, for which he received the bronze medal of the City of Paris. He was also the co-founder and Chairman of International Tramping Tours, (ITT); a movement to promote peace through international understanding. Its aims were to enable groups (British and others) to travel in countries other

Oppposite: Royal Armouries Museum, Leeds, the Hall of Steel.

Above: Ted Happold as a schoolboy, age 12; Ted's parents, Frank and Margaret. They met at Manchester University and are seen here in their graduation robes.

than their own at a minimum cost, tramping and sharing simple accommodation and thereby making intimate contacts with the individuals who constitute the peoples of the world. This dedication to international understanding was reflected in his remarkable relationship with overseas students who entered his department. He followed their careers with great interest, maintained contact, and after his retirement visited many of them offering friendship and advice in equal measure, and as Visiting Professor he advised the Biochemical Department at the University of Ghana at Accra, and the Chinese University in Hong Kong.

When Ted addressed the Royal Designers for Industry at the Royal Society of Arts in 1987, after the election of Lucienne Day as Master of the Royal Designers for Industry, he touched on the problems associated with equality. He said 'Just as women have found it difficult to give expression to and gain recognition for their own particular contributions – their distinction may be a better word – so have engineers. So what can I say that will interest you? Not that I doubt that women are as able as men to lead us. In my own family I learnt this. Great Aunt Helen, who took the first maths degree awarded to a woman at Cambridge, had the best brain in her generation of the family. She was never able to practise, yet served society marvellously. My mother, with her economics and history degrees, who was in at the start of the Youth Hostels Association, founded with my father International Tramping Tours (ITT), took in large numbers of refugee children at the start of the Second World War, then went into local politics out of concern at deprivation in the city, and she ran the planning and housing committees in Leeds for many years.' Ted's sister, Margaret Elfrieda, was a brilliant student and delighted the family by embracing a career in medicine. She was a convinced pacifist who abhorred racism in any form. She applauded her mother's involvement with Second World War refugees, and went one better and adopted four children of mixed race who have bonded into an extraordinary family unit.

It is apparent this intensive dedication to the twin concerns of universal welfare and pacifism and the international peace movement were deeply ingrained in Ted's psyche, as was the family's dedication to learning and scholarship. The high-minded, probing and altruistic polemic that dominated the debate by the family's equally dedicated colleagues and circle of friends was more significant in developing Ted's character than providing a template for his own subsequent idyllic and loving family life. His admiration, love and concern for his parents was apparent throughout their long lives. He also shared their love of village life, and the family colonised the Yorkshire village of Hutton le Hole, first as guests of T Edmund Harvey at his family home at Barmoor and then in family cottages close by. Harvey, a fellow Quaker, befriended the Happolds when they joined the Friends Meeting at Carlton Hill in Leeds. He was a remarkable man, considerably older than the Happolds. He had returned to Leeds after spells at the British Museum and as warden at Chalfont House and Toynbee Hall. He had also been Honorary Treasurer of the Workers Education Association and had three stints in Parliament: 1910–18, 1923–4 and finally as MP for the Combined Universities from 1937 to 1945. He was warden of Swarthmore, the Workers Adult Education college in central Leeds, and he became President of ITT. He was dedicated to young Ted, for whom he became something of a mentor and role model.

Hutton le Hole continued to play a significantly important part in the lives of Ted's own immediate family. He loved the homely village and the Friends Meeting House at Kirkbymoorside with its memories of childhood and his beloved 'Uncle Ted'. He spent the war years at school, first in Leeds at the Grammar School close to home in Headingley. Where his refusal to join the school's army cadet force led to unpleasantness and a pertinent question on pacifism in the House of Commons by T Edmund Harvey, who was influential as an ex-pupil (circa 1887–91) in securing a place for Ted in 1944 at the Quaker Public School at Bootham, York. It was here that he started his 'presidential career', as president of both the Archaeological and the Natural History Society. He also started his collection of medals and scholarships, including the Clayton Travel Scholarship which took him to his first archaeological dig abroad. This was followed by periods working as an ITT tour leader.

In 1949 he entered the University of Leeds Department of Geology where he read physics, maths and geology. He was elected President of the Anthropological Society and in the summer of 1950 joined Professor Ian Richards at an archaeological dig at Hadrian's Wall. He also helped on the Mithraic Temple site at Carrowborough. Then, just prior to his National Service, he took up a travelling scholarship to Cyprus to study medieval military fortifications for four months, based at the Cyprus Museum. In conforming to his Quaker beliefs he registered as a Conscientious Objector and after appearing at a tribunal in Leeds was sent to the Government Experimental Horticultural Station at Stockbridge House in Cawood, Yorkshire, where he worked in a variety of roles – labourer, truck driver, mechanic and drag line operator. However, the pivotal role which started his inevitable vocational change to engineering was his inauguration as chain man setting out the foundations of a large portable greenhouse. When the group constructing it withdrew, Ted offered to complete the task personally – precasting the pulley blocks, digging the base holes and concreting the foundations. The adrenaline rush proved intoxicating and his next move was predictable – an application to Sir Robert McAlpine for employment as a junior engineer. He spent a year on site, setting out, preparing bills of quantities and carrying out site supervision on a multi-storey office building in Leeds.

From 1954 to 1957 he returned to the University of Leeds, this time in the Department of Civil Engineering. He spent his summers on McAlpine sites in Leeds and London, achieved a BSc with honours in Civil Engineering, and was befriended by Sir Basil Spence, then the Hoffman Wood Professor and a regular visitor to the Civil Engineering Department. He advised Ted to make use of his university Moaut Jones Travelling Scholarship and to spend his time in Scandinavia, travelling in Finland, Sweden and Denmark studying modern architectural engineering.

The Happold good fortune and meticulous planning paid off. He intrigued Alvar Aalto, and had a brief spell in his office and an introduction to Finnish hospitality in his holiday home in Saynatsalo. Spence also put in a good word for him with Ove Arup and Partners and in 1957 he started his first formative spell with the partnership, working initially with Povl Ahm on the John Hutton screen frames for Coventry Cathedral. He worked also on a new form of cost analysis, designed the Arup title blocks, published an article on hyperbolic paraboloids and studied architecture at the Regent Street Polytechnic in the evenings.

Like many of us in the late fifties and early sixties, he decided that a spell in the United States was mandatory for his education and once again he chose well. Severud, Elsted and Krueger had achieved an engineering breakthrough

The three people who had such a profound effect on Ted's early life. Opposite: his parents, Margarent and Frank, photographed in 1976; Above: T Edmund Harvey MP, their distinguished friend and mentor (1875–1955).

with the Raleigh Arena in 1952 and this led to a series of beautifully engineered projects with Eero Saarinen, the Yale Hockey Arena and the Dulles Airport Terminal in Washington, both completed in 1958 just prior to Ted's arrival in the Severud office.

It was here that he met Bob Silman who was intrigued by his first vision of Ted as he sat at an adjacent drawing table manipulating a tubular brass instrument that looked like a telescope – pulling and pushing it in and out, twisting it left and right. When he enquired what Ted was doing he promptly replied that he was making very precise calculations with a cylindrical slide rule. This was of course in the days before hand-held calculators or computers. Ted explained that if one were to 'unroll' the surface of the cylindrical slide rule, it would be three times the length of our normal 10-inch straight slide rule, thus producing more accurate answers. Silman, fresh from university, had finished training under a professor who had ingrained in him the notion that precise calculations were meaningless in structural engineering, that it was the order of magnitude that was important. He argued the point with Ted, and quickly lost the argument, but gained a lifelong friend. He and his wife Roberta realised very early in their relationship with Ted that this energetic and unusual Englishman was going to pass through life with 'style', and that he would make the most of every opportunity.

In 1961 Ted returned to Arups, bringing back an even more pronounced work ethic and the organisational and documentation skills of a mainstream American engineering office. The work for this period is well documented elsewhere in the narrative. Suffice it to say his energy and drive were if anything revitalised by America and his workload increased and widened in interest accordingly – buildings with Spence, Hollamby, Renton Howard Wood and Levin, serious pioneering work with Poul Beckman setting up the Arup training school. As Hugh Johnston, the President of CIBSE, observed, Ted's embryonic interest in institutional affairs started at this time, both in his own parent Institution of Structural Engineers and at inter-institutional gatherings organised by the Junior Liaison Organisation, which he later looked back on with particular warmth and affection. It seemed to Johnston that Ted's multi-disciplinary interests were well honed by these experiences and he was sure that they were one among many good reasons why over the years

and at later periods he was in such demand and able to make many vital contributions as Vice-President, President, Master and Chairman of many different and prestigious bodies within the industry and beyond.

The multi-disciplinary talents came into their own in 1967. That really was Ted's year: the competition win at Riyadh with Trevor Dannatt, the fateful meeting with Frei Otto and Rolf Gutbrod on the same trip which prefaced such a long and rewarding partnership. The year also provided Ted with what his friends and colleagues always felt was his richest prize, his wife Eve. She gave immediate stability to his life, provided a wonderful home, a continuous loving relationship and after a short time two fine sons, Matthew and Tom. Even the wedding kept its links with construction – the reception was held in the recently completed Hollamby/Happold extension to the West Norwood Library, and from that moment on the real partnership of life began. As Jill Loveless, one of the family's closest friends, remarked 'Ted was never happier than when surrounded by family and friends. He loved and was loved by many people, but Eve, Matthew and Tom were the centre of his life – the still point of a turning world as T S Eliot put it. One of Ted's greatest qualities was his enjoyment of life. He was always enthusiastic, whether the occasion was a visit to the Bath Festival, a trip to the Chichester Theatre, or a rather chilly picnic on the Sussex Downs. He loved entertaining, and was a superb host.' (Many will remember his fiftieth birthday party at Laycock Abbey, the Happolds wonderful Silver Wedding party and scores of happy occasions, always brilliantly organised by Eve.)

It was a pleasure to be part of the Happold family picture. On each occasion hospitality and harmony ran hand in hand. Eve helped to uncork the genie out of the bottle and Ted's effervescence and freedom of thought were encouraged and nurtured in a relationship that was always Christian, creative and concerned. They fought the battles together, enjoyed the triumphs, supported each other in adversity and travelled together whenever they could. Eve embraced Ted's crazy lifestyle, cemented all his relationships with his colleagues and collaborators and kept an equilibrium on the home front which gave Ted real contentment and happiness and Matthew and Tom a totally stable and stimulating home life. This stimulation was further enhanced by Ted's stated ambition 'that his favourite pursuit was to travel with his family experiencing together new and interesting locations'

Opposite, clockwise from top left: Young father Ted with the two boys, Tom and Matthew, at Hutton le Hole 1971; The family on the IStructE presidential tour in 1987 at Agra; Bottom: Ted's cylindrical slide rule that so intrigued Bob Silman.

Above: The sleepy village of Hutton le Hole; Ted and Tom equipped for camping, May 1984.

– ranging from Death Valley to 'Scottish Baronial Banff' in Alberta. These coupled with annual skiing holidays at Lech, chosen for continuity and local knowledge, but always approached by interesting routes! However, the trip that finally encapsulated all his ambitions for the family was his journey as President of the Institute of Structural Engineers in 1987. The 'official' destination was the Indian sub-continent but the approach was vintage Ted – a start in Calgary (shades of the Saddledome and 58°N), then via the Canadian Pacific railroad to Vancouver, then on to Tokyo (one of the few places he had never been), then to Hong Kong and Singapore, combining institutional business with family adventures. After three weeks en route they reached Bombay and with unbelievable rapidity Karachi, Islamabad, Rawalpindi, Lahore and Delhi, all involved in the official visit and festooned with meetings, lectures and speeches. Ted tagged on a family holiday taking in Agra, Fathepur Sikri and Jaipur, the latter featuring a vintage Stanley and Livingstone re-enactment. Quite by chance I was visiting Rajasthan (unaware of the Happold family whereabouts) and was gingerly dismounting from a rather languid elephant after ascending to the courtyard of the Amber Palace in Jaipur when I heard an unmistakable cackle of throaty laughter and looked up to see the family Happold en masse. A most unlikely but joyful coincidence which finished up with a life threatening meal in the Rambagh Palace that evening.

In 1976 the University of Bath offered Ted the Chair of Building Engineering with the opportunity to develop a joint school of architecture and engineering, which he carried off brilliantly. Simultaneously he moved home from London to Bath and with old alumni from Structures 3 – Ian Liddell, Michael Dickson, Peter Buckthorp, Terry Ealey, Rod MacDonald, John Morrison and John Reid and formed Buro Happold. It was a collective in the real sense. The Liddells were housed first, and tended to play host to all the partners and staff. An office was started from scratch in Gay Street, and in two decades the partnership became a truly international practice with wide-ranging and sophisticated design skills. Ted's way was always to fashion a collective spirit, which now seems second nature to his émigré friends from Structures 3, who are dedicated to the standards he demanded and continue the search for that perceptive and inventive streak which illuminated all his engineering solutions.

In discussing Ted's qualities as an engineer, it is interesting that the negatives have inevitably come from people who have not worked with him intimately ... Purely an entrepreneur ... He was not safe ... An opportunist ... A conceptualist ... A collector of talent ... Perhaps all these are true, but let us place them in the context of statements by some of his closest colleagues and collaborators. First, Frei Otto;

'Within the last few years I have lost four of my friends – Bucky Fuller, Ove Arup, Peter Rice and now Ted. In Stuttgart we still remember Bucky dancing on a table at my Institute and singing the Bauhaus song after a long day working on biological structure. Ove Arup was smiling and listening, and next day we visited our common work known as the "Miracle of Mannheim". Among these four gentlemen Ted was, in my opinion, the greatest. He was so human, so understanding, sympathetic and appreciative. Ted had a good feeling and taste in arts. He was a man of science living in the world of history and arts. I think he was not only a scientist who stands behind the architect but also a scientific architect.'

The first project that Buro Happold carried out with Ted Cullinan involved the buildings at Bedfont Lakes, carried out in conjunction with Michael Hopkins. As Cullinan said: 'Before then I'd only known Ted at parties and sometimes his parties, which he gave with all his energy and wit and which sometimes achieved splendour, like his fiftieth at Laycock Abbey. Since then we've worked together whenever possible, and with Ted involved, it was sometimes hard to distinguish the work from the parties. For he had absolutely no pompousness or self-importance or pseudo seriousness or pedantry or pernickityness or any of the other "I'm at work" characteristics which you find in "I'm at a meeting" people. He was always Ted, properly Ted; work or play.'

Bryan Harris, his colleague at the University of Bath and Professor of Materials Science, felt that 'Ted had a fine vision of creating an environment in which student architects and engineers could learn together, could learn in common some of the technical and aesthetic aspects of their subjects, could learn a little of each other's languages, could in this way come to understand each other's problems and perhaps respect each other's professionalism rather more than was, perhaps even still is, the norm. Ted believed that "a world that sees art and engineering as divided is not seeing the world as a whole". It seems a reasonable notion, but there was opposition, mainly from lesser people with axes to grind, but Ted knew what he was talking about. He had himself the mathematical skills of the engineer and the aesthetic sense of an artist and he was a great persuader. (I have an image already of celestial architects being persuaded to abandon their Palladian and Gothic forms in favour of grid shells.) Ted succeeded in building something unique which has brought appreciation and distinction to the University of Bath and though perhaps the institution does not yet fully understand the significance of Ted's contribution to its life, its work and its reputation, the world outside is in no doubt.'

It is perhaps his closest colleagues and fellow engineers in Structures 3 who have the measure of the man and his work as an engineer.

Michael Dickson is in no doubt about Ted's passionate belief in the importance of the creative role of engineers: 'The many projects which bear his name are a fine testimony to this. As a designer and engineer he had a rare combination of artistic judgement and scientific knowledge which enabled him to put his finger at the centre of a design proposition knowing that the composition could work elegantly, completely and economically. That is a rare talent.'

Ian Liddell noted that 'Ted, in talking about his view of the engineer's role, argues that the "romantic mode" tends only to develop existing forms whereas the "classical mode" is capable of producing originality.' Ted wrote 'Most engineers would see themselves falling into the latter category. Their craft is intensely creative; at its best it is art in that it extends people's vision of what is possible and gives them new insights. But the aesthetic produced is bare (meaning unfinished). It may not have been seen before and is more likely to relate to a natural rather than historical precedent. This is what I mean by engineering design as a technological idea as distinct from a visual style or fashion.'

'Many of Ted's buildings had this quality of a bare aesthetic which accompanied technological advances. One could think of Mannheim, or the structure of the Diplomatic Club, and perhaps most obviously in Centre Pompidou, which he did so much to get off the ground.'

Finally, Terry Ealey, speaking at Ted's memorial service, remarked on 'Ted's unusual combination of talent, drive, humanity and charismatic personality,

Opposite, top: Ted's portrait as Chairman of the Trustees of the Theatre Royal, Bath, 1993; Bottom: Ted and the boys on a family holiday in the American Southwest.

Top: Ted and Eve at Tom's degree ceremony, July 1995; Above: Matthew and Tom and a rather incapacitated Ted in Hong Kong, summer 1982.

which were the foundation of his life and achievement. Ted knew a thing or two about foundations having taken degrees in both geology and civil engineering at Leeds University. However, his passion was for building superstructures. He was a builder of buildings, but also a builder in a much wider sense and he built with his friends. His friends' lives have often been better spanned, better formed and more innovative as a result of knowing him and working with him.'

Even to the end Ted felt the need to focus on his options. In a short piece outlining 'Thoughts on what I can still do with my life', the opening paragraph set the tone:

'In a sense the choice is quite a stark one – either my already restricted life at home with limited contact with others becomes more constrained until I die in a year or so, or I am fortunate to get a compatible heart and there is no reason why I could not live as long as my parents, both of whom died near their ninetieth year.

I suppose one realises that the art of living must be studied and considered, as must every art. It calls for imagination, so that every advance, every change, is not merely different but a creative act. Achievement calls for self-discipline.

I do not think I fear death – after all I have been very fortunate in having had an intensely fascinating life – though I am concerned at the problems it would give my wife by my not sharing in the tasks (and for that matter the pleasures) of life and parenthood. But I do feel that there is a lot still I can and want to do and I think I am angry now (though I am not sure who with!) at the possibility of not being able to do them. Who knows – perhaps this is sent to make me examine how much I fail to practise what I profess.

Civil and building engineering, like medicine, is a very broad subject. Starting from being able to define and articulate human needs it requires a talent for creating economic and effective solutions, knowledge and rigour in using scientific and other methods to test and evaluate those solutions, a wide-ranging ability to communicate, and management skills for construction

and operation. Scale and complexity is such that it is a group activity with a high social content and a high social purpose. It has always seemed to me very satisfying because it has within it almost every aspect of life. One can study its history, one can take part in its internal politics, one can be interested in the development of its science – climatology, materials science, numerical analysis and so on – the pleasure of designing environmentally responsible buildings and the social side of interrelationships – and all hopefully helping society and providing one with a living.'

Happy Birthdays

He followed this preamble with a summary of design challenges, possible committee work, research initiatives, the streamlining of practice management, and setting up a charitable educational trust. Alas, he was not given the time, but the little 'General' really fitted ten careers into his life and his messianic zeal in pursuit of problem solving brought on regular bouts of amazement, hysteria, exasperation and admiration in equal measure to both his partners at Buro Happold and those of us in architecture who sought his involvement.

At his memorial service the sensitive eulogy by his son Matthew, speaking on behalf of his mother Eve and his brother Tom, touched everyone deeply and indicated what a very special man Ted Happold was. He said :

'In truth, a few minutes is simply insufficient to give any proper acknowledgement of his many qualities and achievements. So I'll just say that I loved him very much, I respected him deeply and I am convinced that my memories of him will always be with me to encourage me and to guide me in the decisions I will have to take throughout my life.

William Penn, the Quaker founder of Pennsylvania and a writer of whom my father was very fond wrote in 1682 that: "True godliness don't turn men out of the world, but enables them to live better in it, and excites their endeavours to mend it: not to hide their candle under a bushel, but set it on a table in a candlestick. I believe that this was true of my father and that the light of his example will continue to shine. Though I mourn his loss deeply, for this, and for many other of his gifts, I give thanks to God for the life of my father.'

Opposite, clockwise from top left: Ted assessing a competition in Hangzhou; Ted and Eve enjoying Chinese Gardens; Ted sightseeing on Island on the Lake in Hangzhou on the same trip, 1993.

Top left: Ted and Eve at 1985 Bath University Conference; Top right: Ted in party mode; below: A typical Ted greeting – as a 130-year-old ackermann.

When we collaborated together on a book to celebrate the 150th anniversary of the Royal College of Art called, prophetically, *The Great Engineers*, I felt it appropriate to describe the unique quality of today's creative engineers by summarising Happold's personality as entrepreneurial, inventive, analytical, literate and lively. As an epitaph I will settle for that. He is sorely missed at every level. His family, his friends and his colleagues miss the richness of his personality, his wisdom, humour and spirituality. His was a singular life, much fuller and more productive than seems possible – and what he gave to his family, his friends and his colleagues was unique. His last resting place, unlike Rupert Brooke's is not a foreign field but close to his roots in the beautiful garden of the Friends Meeting House he loved so much in Kirkbymoorside. "So he passed over, and the trumpets sounded for him on the other side.'

Opposite, clockwise from top left: 'Sir Ted' with the family at Buckingham Palace, February 1994; Ted and Eve at Alison Jones's wedding, July 1994; Ted, Matthew and Eve return from skiing holiday in Lech, Easter 1984; 'Teddy Mecury' LA. The final holiday with the Silmans at Dove Cottage, May 1995; student competition crit 1987, Ted in full flow with Derek Walker to his right and Kit Allsopp to his left.

Above, from top left: The Buro Happold office at Camden Mill, Bath; An early photograph of Ted with Frei Otto in 1974 on the site for the Mannheim Garden Festival; The eco-conscious house designed for the Happolds by Frei Otto in 1995; Friends Meeting House at Blackheath on which Ted worked with Trevor Dannatt.

The Practice of Engineering

I first met Ted Happold on my return from Cornell in 1968 when I joined Structures 3 at Ove Arup and Partners. Ted immediately dispatched me for ten weeks' indoctrination at the Arup Graduate Training School. Apart from experiencing the energy he imparted to everyone in Structures 3, I didn't get to know him properly until he drove me in his ageing but classic Jaguar for a meeting with Frei Otto in 1972. In the following twenty-five years as his partner and friend I have been inspired by his confidence and support and have learned to tackle, often directly with his advice, a wide range of construction projects. I hope I can capture the dynamic of both the range of work he accomplished in his life – its elegance, its technical sophistication – and the great fun we all had in working with him.

Much of Ted Happold's professional life was driven by his belief that good creative engineering can stimulate good architecture. He believed in the joint working of architects and engineers. As an engineer his inclination was to pursue efficiency and economy while conscious of the artistic, social, symbolic and intellectual issues of our built environment. He understood the need to reduce architecture and science to the practicalities required of construction, where meticulous engineering and a sense of scale and ingenuity of detail add to the nature of the architectural solution. His Quaker beliefs shaped his view that the engineer's role is to improve the public domain and that work is to be enjoyed.

With reference to history – Smeaton et al – he claimed that what people actually buy, particularly in developing countries, is the engineering – the functional efficiency. In this respect he was critical of the UK building industry and its resistance to change, but he was the last person to belittle the skills and contributions of the many professions that comprise the construction industry. His views, in later years, led to his work on the establishment of the Construction Industry Council. He was fortunate in his formative years after university to learn the skills of design and detailing while working almost directly for Ove Arup with Povl Ahm and to have worked in the New York offices of Severud, Elsted and Krueger. Here he learned the importance in engineering of vision, courage and discipline. That very courage was the singular ingredient that be brought as an engineer to the many architects with whom he worked during his life. Robert Silman a colleague there, now Senior Partner of Silman Associates, recalls 'For Ted, anything was possible and over the years he succeeded in convincing me that I could achieve whatever goals I set for myself'. Many who worked with Ted subsequently would say the same.

When Ted returned to London in 1961 he worked on further buildings of Sir Basil Spence, initially under Povl Ahm and later as project leader in Ove Arup and Partners. He was an engineer who did not automatically choose the easy route or the stock answer to a problem. Buildings of the complexity of the British Embassy in Rome testify to that. In 1967 he became executive partner of Structures 3, an appointment which reflected Ted's enthusiasm for good engineering and heralded a rush of imaginative and successful projects. First, Trevor Dannatt and Ted Happold won the competition instigated by King Faisal for the Conference Centre and Hotel in Riyadh. Ian Liddell, Ted's right-hand man for many subsequent years and a founding partner of Buro Happold, returned to Arups to carry out the engineering work for Riyadh with

Opposite, clockwise from top left: Kafesses at Mecca Conference Centre (Rolf Gutbrod, Frei Otto and Ove Arup & Partners); Coventry Cathedral front with Epstein sculpture (Sir Basil Spence and Ove Arup & Partners); Mannheim, Bundesgartenschau (Carlfried Mutschler and Partner, Frei Otto, and Ove Arup & Partners); Escalators at Beaubourg, Paris (Piano Rogers and Ove Arup & Partners).

Above: Ove Arup and Ted Happold at Ove's 80th birthday party; The Beaubourg gerberettes.

Michael Dickson, the present chairman of Buro Happold and one of Ted Happold's closest friends and colleagues wrote this piece to record Ted Happold's singular contribution to the practice of engineering

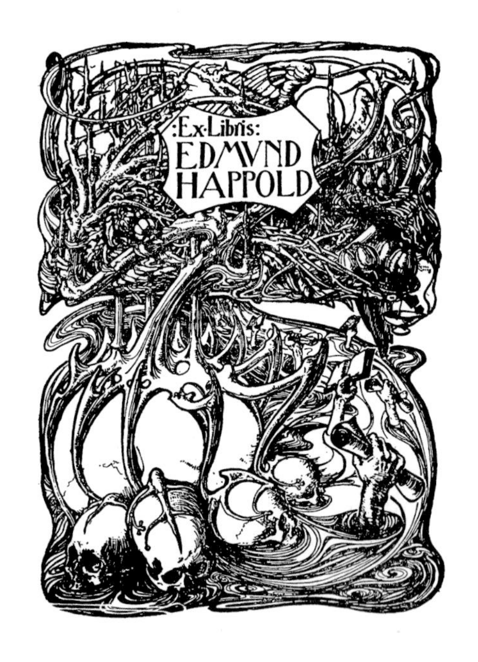

Ted. The structure of the Riyadh Foyer and of the column support to the space-frame roof to the Conference Hall both illustrate how structural design – here for thermal movement – can be articulated to the benefit of the architecture.

When he was in Riyadh with Trevor Dannatt Ted met fellow competitors Professors Rolf Gutbrod and Frei Otto, who were impressed with the multiple skills Ted had to offer. As a result, Arups were commissioned to engineer their project for the Mecca Conference Centre. Computing the geometry of the complex fan-shaped roof plan under its own weight was no easy task in the late sixties. They had to use the main-frame computer at Manchester University. In its use of the lightest possible 'heavyweight' insulation and external reflection of energy, the constructional design recognised the interaction between the structure of hanging cables and stiffening structural angles with the carapace of the building itself.

In the late sixties and early seventies work being undertaken by the Structures 3 team had an extraordinary range. From architects Renton Howard Wood Levin came a new generation of British theatres: the Crucible Theatre in Sheffield with its full-thrust stage and then the Arts Theatre Complex on the emerging campus of Warwick University. The latter included a single rake 900-seat auditorium in front of a semi-thrust stage and a small studio theatre, both useful as an experiment in form for theatre consultant John Bury. Bury was later to help Lasdun with the auditoria for the National Theatre and be Ted's fellow director in the modernisation of the Theatre Royal, Bath, in the nineties. With Rentons too there was the refurbishment of St Katherine's Dock by Tower Bridge for Taylor Woodrow Properties. This project developed Ted's interest in the 'appraisal of renovation of structures', which led to his chairing of the Institution of Structural Engineers' Appraisal of Structures Group where he was helped by, among others, Professor Bill Biggs, Poul Beckman and Bob Milne, the group's Technical Secretary.

Ted's friendships in Arabia led to Arups' appointment to master plan the University of Riyadh with architect Karl Schwanzer. Rod Macdonald, later to be Ted's partner, was closely involved in this project and the master planning and design of Qatar University with Dr Kamal Khafrawi and Renton Howard Wood. An interesting architectural device of this scheme were the environmental towers of wind. The physical understanding of how these towers actually worked brought Max Fordham, the building services designer, on his bicycle to Ted's house in Gloucester Crescent late one night with the words 'Ted, Ted, I've got it!'. Such enthusiasms, stemming from a lifelong comradeship in engineering, were common in all Ted's initiatives and were part of the formal incubation of his interest in the design of the environment.

Ted's involvement in Beaubourg was certainly a watershed in his career. On the one hand, Ted's creative touch as an engineer can be seen in the design concept of castings, water-filled tubes and long spans for the steel structure for Centre Pompidou. In the book *State of British Architecture* Sutherland Lyall wrote … Designed by Piano and Rogers and engineer Ted Happold, the Pompidou Centre is a big shed inside out, six large clear floors enclosed by a glass skin with all the structure and services runs and most of the circulation on the outside.' On the other, Ted's involvement in both the competition and the subsequent negotiation with the French authorities is regarded by Rogers and Renzo Piano as pivotal to the successful construction of this seventies icon.

In 1973 Ted returned full time to Structures 3 to rejoin Michael Barclay and others. He started a lightweight structures laboratory as a corresponding organisation to Frei Otto's Institut für Leichte Flächentragwerke in Stuttgart. This led to designs for a number of lightweight coverings, including the tent at Dyce (with DRU for the BP oilfields), opened by the Queen in 1975. This 80m x 40m humped tent in sewn cotton polyester canvas had to be designed and constructed in ten weeks as a 1,000-person auditorium. As a result of this initiative this small group within Arups was responsible in due course for the emergence – and full architectural acceptance – of lightweight structures for permanent use in the UK.

Of all the projects that he executed with Frei Otto, the gridshell for Mannheim, Bundesgartenshau (1975) with architect Carlfried Mutschler best portrays Ted's singular skills in planning a design and its execution. The paper on this project presented with Ian Liddell at the Institution of Structural Engineers pinpoints in a particular way Ted's knowledge of the design process and the use of structural materials, coupled with his confidence that disparate teams can – with patience, hard work and skill – solve the almost unsolvable.

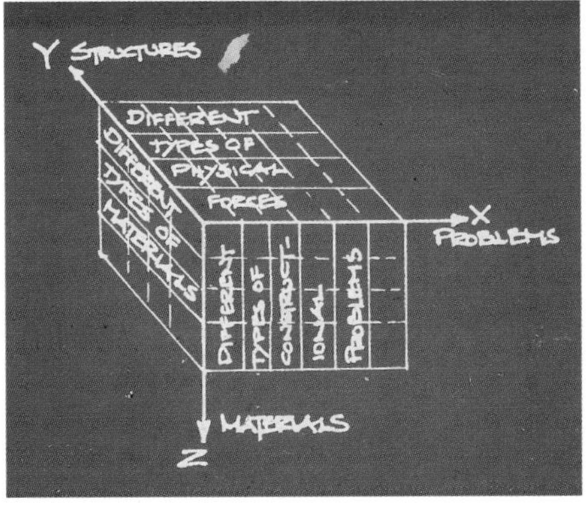

It was no surprise – considering his family background, his early teaching experience both at Cambridge and within the Arup organisation – that his long-term aim was to combine practice with teaching and research. In 1976 he took up the position of Professor of Building Engineering (and alternating head, with Michael Brawne) at the joint School of Architecture and Engineering at the University of Bath. At the same time Rolf Gutbrod offered him elements of the structural and civil engineering design for the Kocommas project (King's Office, Council of Ministers, Majlis al Shura) with Ove Arup and Partners. So on the 1st of May 1976 he was able to put up his plate at 14 Gay Street, Bath, as Senior Partner of Buro Happold.

In his eighteen years at Bath University, Ted taught at least two generations of young architects and engineers the principles of mutual professional respect. He set up the Centre for Window and Cladding Technology and the Wolfson Air Supported Research Group. The latter corresponded to the SFB 64 Widespan Group in Stuttgart and it set about assembling all the pieces of the scientific jigsaw puzzle of techniques and technologies needed for lightweight tensioned structures. This work remains to this day a 'vade-mecum' of engineering research, encompassing as it does all the disciplines – materials research and testing, design method, economics, analysis and forensic observation, wind tunnel work and environmental design – it was true engineering research. The work led to the study, with Arni Fullerton, for a 35-acre covering in northern Alberta for a permanent village of 2,000 people (with houses, schools, parks, etc) in a region free of frost for only ten weeks of the year. The design used the economies of an air supported structure while creating a form orientated towards the low angled winter sun and employing the emerging technology of using ETFE foil cushions as roof cladding.

Throughout his life, lightweight structures remained Ted's passion. With Theo Crosby in 1976, Ted, Ian Liddell and Eddie Pugh designed the tents for the Challenge to British Genius. With Frei Otto and Rolf Gutbrod there was the 120m x 90m giant nomadic tent form in Jeddah constructed on the basis of Frei Otto's soap bubble model as a prestressed cable net. The materials research involved led to the development by Sarna of a special PVC-coated polyester grid weave for the membrane covering.

Opposite, from top: Ted's book imprint adapted from one used by T Edmund Harvey; Ted with Frei Otto and Walter Bird (Birdair); Energy-catching structure at Badensweiler with Rolf Gutbrod; Rolf Gutbrod and Frei Otto in Arabia.

Above: Ted at Hooke Park on a site visit; One of Ted's favourite diagrams, 'Structures, Materials and Problems'

Interaction between Ted's practice and research group produced the design in stainless steel grid weave for the 90m x 40m free form aviary at Munich Zoo, with Jorg Gribl and Frei Otto. Designed as a cloud, with a form to accommodate the habitation and flight patterns of bird life, it was erected to its doubly curved form with pantograph mast heads from a flat sheet of mesh. The design required extensive static and fatigue testing of the spring wires at Bath University in order to gain acceptance by the Proof Engineer in Munich, Dr Kupfer.

Another notable lightweight structure carried out by Buro Happold under Ted's direction – with Ian Liddell, Mike Cook and Eddie Pugh – was the unusual covering to the atrium of the Imagination Building in Store Street, London, with the late Ron Herron. Ted was delighted that the engineering contribution helped this project to gain the British Construction Industry Award in 1989.

In USA, work with Nicholas Goldsmith and Todd Dalland of FTL led to a number of outstanding tensile projects with finely defined engineering forms, including the Baltimore Music Festival Canopy, the Boston Harbour Lights and the AT&T hospitality tents for the Atlanta Olympics in 1996.

Ted had an unusual breadth of vision for an engineer and believed in the importance of undertaking competitions as a way of articulating emerging ideas and moving forward. The designs for Pompidou, the energy-catching structures for Badensweiler with Rolf Gutbrod, the Mosque in Brussels with Theo Crosby, Vauxhall Cross with Kit Allsop and Andrew Sebire, the Diplomatic Club in Riyadh with Frei Otto and Sprankle Lynd & Sprague, for the Leeds Playhouse and the Wycombe Arts Centre with Derek Walker all demonstrate this.

With Christoph Ingenhoven (IOP) and Frei Otto, the Green Tower competition for the Frankfurt Commerz Bank shows the importance of integrating engineering performance with innovative cladding elements to achieve building efficiency. The proposition had a circular, externally braced, 44-storey tower only 40m in diameter, with a quatrefoil office plan cantilevered from four columns so that every office, whether facing north or south, faced on to an atrium within the triple-glazed façades. This cladding design was taken forward by IOP into the 30-storey 120m-high circular office tower for RWE in Essen. The triple-glazed façade clad a reinforced concrete structure where the concept was to use a rib form of concrete for additional thermal mass which could house purpose-formed chilled beams, so-called 'surf boards', complete with integrated sprinkler and lighting facilities.

No essay on Ted can be complete without reference to his thirty years of working on projects in Arabia with so many talented and able people from these countries. In Kuwait, he undertook the Islamic Medical Centre and many other projects with his friend Dr Khalid Al Marzook. In Riyadh with Omrania and Frei Otto there was the remarkable Diplomatic Club for Dr Mohammed Al Sheik. The controlled environment within the curving, massive stone and concrete walls and lightweight tensile shells houses the accommodation and surrounds the garden with its stunning stainless steel cable and stained glass Heart Tent. With Abdul Wahib El Wakil the practice collaborated on the rebuilding of the domes and minarets of the Quba Mosque in Medina in traditional hollow clay blocks, without using form work. The courtyard was subsequently covered by an unfurling Toldo roof designed by

Dr Bodo Rasch of Stuttgart. This project was followed by the folding umbrellas for the Prophets Mosque in Medina. Then, working with Ted's partner Terry Ealey and architect Rasem Badran, the team designed and built the Grand Mosque and Justice Palace for the Arriyadh Development Authority following traditional Najd design but incorporating within the head of the modern two-sided column beams air conditioning and lighting. On this project the client's representative was Ahmed Salloum, and in 1995 the Grand Mosque received an Aga Khan Award for architecture in the Islamic world.

The full designs for the new 250m office tower, hotel, retail and banqueting complex in Riyadh for the King Faisal Foundation (by a joint venture of Sir Norman Foster & Partners and Buro Happold) were completed in 1996. Ted's trip to Jeddah in 1995 with the Buro Happold Bath team to undertake discussions with Prince Khalid and Prince Bandar and their advisors was the last overseas trip he was able to make.

In the UK, Ted Happold had a long standing Yorkshire friendship with his friend, co-author of this book, Derek Walker. I can remember Ted's proposal in 1975 (derived from his knowledge of the hanging plywood roofs he saw at the Lausanne exhibition in 1964) for thin stainless steel hanging sheet roofs over the City Club in Milton Keynes where Derek was Chief Architect and Planner. In Hong Kong their association led to the complete redevelopment and construction to Olympic standards of a sports centre and swimming pool in a 30-acre landscaped park, and a sophisticated masterplan proposal for the third Hong Kong University. In Leeds, they had to wait until the 1996 opening of the new Royal Armouries for a completed project in the UK. The Armouries, with its integrated services and structure and its remarkable Hall of Steel in stainless steel and glass, was engineered by Rod Macdonald and Buro Happold Leeds.

Interaction between the practice and engineering research at the University of Bath was the basis of Ted's ability to innovate. Right up until the day before his death, 12 January 1996, Ted was working on the use of round softwood thinnings for construction in the UK – a project started with Richard Burton, Frei Otto and John Makepeace for the School for New Woodland Industries. Following a remarkable prototype house, where the central idea for the use of thinnings as hanging bundles of parallel cellulose was enabled by the adaptation and testing of a glued two-part epoxy joint, there was the 15m x 45m clear-span workshop for the school with bent roundwood arches. Recently, with Ted Cullinan Architects, the new eight-person Westminster Lodge is the product of joint design and research by Buro Happold and Bryan Harris and Stephen Ring at the University of Bath.

Ted had a great many friends and enormous energy, and was widely knowledgeable in construction. He was a great inspiration to team working and was great fun to be with. He was one of those people whose lives result in a perceptible shift in the human condition, so he was justly proud of his appointment as Master of the Royal Designers for Industry in 1992, and it was significant that his extraordinary work in Germany should be honoured by the granting posthumously of an honorary doctorate from Braunschweig University. His inspiration lives on at the University of Bath where (Michael Barnes has succeeded him as Professor of Civil Engineering) and in the various design offices of Buro Happold, through his many friends and colleagues.

Opposite, from top: Interior of the staff house at Hooke Park where, apart from the 'A'-shaped column most of the members are in tension (ABK, Buro Happold); The wind towers at the University of Qatar (Kamal Khafrawi, Renton Howard Wood Levin, Ove Arup & Partners); The Munich Aviary (Jorg Gribl, Frei Otto, Buro Happold).

Above: Early partnership portrait, Buro Happold, Bath (Ted Happold foreground right); A cartoon of Ted carried out at Ove Arups.

A Personal Perception of Engineering

This piece was written by Ted Happold for the publication I edited for the Great Engineers Exhibition which celebrated the 150th anniversary of the foundation of the Royal College of Art in 1987. It echoes a constant theme in Happold's singular stance on the art of the engineer – his roots, his lateral vision and his dedication to the creative team as a force in construction. DW

Everyone knows the harshness of nature; how mistakes become extinct. Yet the unique characteristic of man is that his reason and imagination have enabled him to develop and adapt to his environment. And history shows how technological development has liberated him.

At first it was the discovery of a type of wheat which gave an abundant yield, achieved by ploughing, sowing and reaping, which led to the possibility of settlement and the building of permanent structures. Today there is the exploration of the deep – the conquest of space.

Yet in response there has been continuous concern at such advances. Fear of change. Fear of consequences. Technology, especially since the time of the industrial revolution, has been seen as a dehumanising force to be resisted.

Art has been seen as the civilising counterbalance to these advances. Art is seen as expressing the individuality of man and promoting cultural evolution. It took until 1836 for a government report on Arts and Manufacturing to recommend the formation of a School of Design in which 'the direct application of the arts for manufacturing should be deemed an essential element'.

But technology (defined by Galbraith as 'the systematic application of scientific or other organised knowledge to practical tasks') – or engineering if you prefer that word – obviously cannot in itself be a bad thing; it is how we use it. And a world which sees art and engineering as divided is not seeing the world as a whole. These days often only those people with an arts training are said to be creative. But, if the truth be told, it is technology that is creative because it gives new opportunities. Historic ideas of art and culture can entrap. It is technology that frees the scene.

Throughout history there has been a succession of turning points, achievements by engineers which represent a new conception of nature. This narrative is about these turning points, which express why engineering can be so intensely satisfying because it is, at its best, an art grounded in social responsibility.

Industrial archaeology – ie the history of engineering – is now culturally acceptable and certainly influences modern fashion. Yet as an engineer designer I represent an approach to design whose roots are not dependent on visual precedents. I am referring to engineering design as a technological idea, with its own aesthetic.

The engineering profession as we know it today developed to serve the nonconformist industrialists of the eighteenth century, who were the midwives to the industrial revolution. These men, predominantly Quakers, were barred by dint of their non-conformity from the established universities and professions. They found creative opportunities within their limited possibilities by turning

Opposite: The Heart Tent, Diplomatic Club, Riyadh (Omrania, Frei Otto, Buro Happold) with stained glass design by Bettina Otto.

Above, from top: John Smeaton, a fellow engineer from Leeds and Ted's one great hero, the first professional engineer; James Brindley, the eminent canal builder. His Bridgewater Canal, completed 1769, popularised the use of puddled clay.

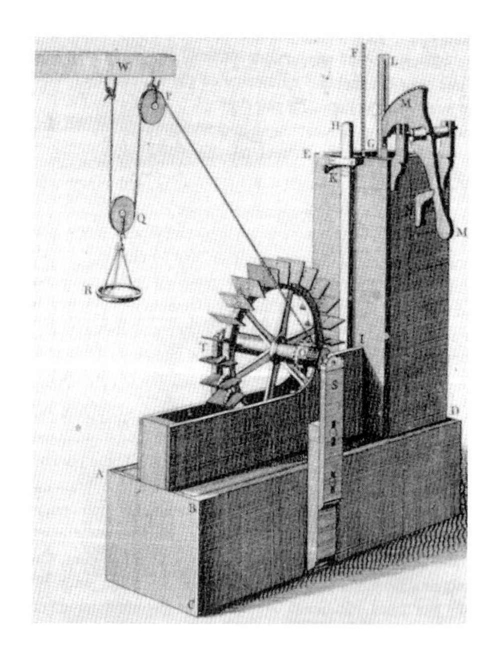

to inventive industry. And because they believed in the equality of mankind guided by individual conscience, they backed humanistic management. They had very broad long-term aims. They coped with persecution by forming close bonds with their fellow industrialists and these inter-relationships and the pooling of ideas and information facilitated the development of the industrial revolution.

The founders of our profession were creative mechanics. In 1712 Newcomen erected his first steam engine. In 1759 Smeaton carried out a classic study on water power. In the forty years that he worked as a consulting engineer, Smeaton regularly used the word 'professional' to describe himself. Each morning he was employed on a time basis to consider problems and design schemes. But the fact that he saw his scientific studies as the basis of his work is described by his daughter, Mary, who wrote 'his afternoons were regularly occupied by practical experiments, or some other branch of mechanics'. This interest in the scientific study of the sources of power in nature, together with the performance of materials, represents one aspect of a technologist's body of knowledge. The other aspect is in the development of construction methods; the organisation of work. Some of the achievements of the other accredited founder of the civil engineering profession in Britain, James Brindley, well illustrate this. In the construction of the Bridgewater Canal (completed 1769) he popularised the use of 'puddled clay': mixing together readily available sand and clay and getting his workmen to tramp it into the bottom of the canal with their boots to provide an impervious yet flexible lining.

Engineering technology really took off in 1760 when two foremen at a Quaker ironworks in Coalbrookdale produced cheap iron using coal, not wood, as fuel. It was the beginning of the era of making materials with creative possibilities using non-renewable resources, such as metals and fossil fuels.

In 1779 the world's first iron bridge at Coalbrookdale was complete. In 1801 the first steam carriage was built by Richard Trevithick. In 1826 Telford's great suspension bridge over the Menai was completed. George and Robert Stephenson built the first effective steam railway and I K Brunel's steam ships bridged the Atlantic. Technology transfer started early – every American knows the portrait of Whistler's mother – but his father came over to Britain to learn railway engineering from the Stephensons and not only returned to pioneer railways in America, but also went to Russia and started the construction of the trans-Siberian railway. These engineers started the staggering development of technologies which have in the last 140 years so changed the world.

As a structural engineer, or technologist, I acknowledge that my way of thinking about the world falls into what Persig designates the 'classical mode'. You may remember that in his book *Zen and the Art of Motor Cycle Maintenance* he describes those who see the world primarily in terms of immediate appearance as thinking in the 'romantic mode' and those who see the world primarily in terms of underlying form as thinking in the 'classical mode'.

His interesting example is of Mark Twain, who wrote in lyrical terms about the Mississippi River until he went to train as a river boat pilot. He gained a deeper understanding of the river through learning its science, but in the process it lost its original magic. Persig argues that the romantic mode tends only to develop existing forms, whereas the classical mode is capable of producing originality.

Most engineers would see themselves as falling into the latter category. Their craft is intensely creative; at its best it is art, in that it extends people's vision of what is possible and gives them new insights. But the aesthetic produced is 'bare', it may not have been seen before, and it is more likely to relate to a natural, rather than historical, precedent. This is what I mean by engineering design as a technological idea as distinct from a visual style or fashion.

Perhaps the best-known example of this is the Crystal Palace, produced not as an 'art object' but because it worked. Prefabrication and organisation of plant and labour, together with the significance of iron and glass as building materials, were demonstrated to the world. But to the art/architecture establishment of the time, notably Ruskin, the Crystal Palace was anathema. He advocated a return to the style of Gothic Mediaevalism. It is also interesting that although fire-proofed cast-iron structures were produced only two years after the erection of the Ironbridge at Coalbrookdale, it was a quarter of a century before they were used in architect designed buildings.

Certainly in Britain there is a belief that technology must be tamed and controlled, largely by imposing standards of visual beauty which comply with criteria set up by those who have studied the arts.

Sixty-five years ago Roger Fry, amongst others, was complaining that the British were in love with ancient art and too little interested in the work of their contemporaries. I sometimes wonder whether we were not almost fatally damaged by the refugees from Hitler's Europe. Those who were gifted designers, frustrated by the reactionary environment here, went on to the USA and revolutionised design there. The art critics and historians, discovering our conservative, class-rich environment, stayed on to become famous and powerful in our establishment.

But technology is about change. It is concerned with the development of useful objects or processes which change our lives. It does this in response to people's aspirations or is restrained by people's fears; in this it relates to the arts. What it does must obey the laws of nature, which is why it uses science to examine behaviour. Technology is the making of things while science is the explaining. So the roots of engineering are in nature.

Everything in the built environment has been achieved by technology. Every single man-made object in the world is the product of technology and traditionally the modern built environment divides into structures and machines.

The relationship between structures and machines is extremely interesting and there are many aspects they have in common. The most efficient use of materials, or perhaps energy, is a major one. Yet the energy in structures, which are seen as essentially permanent, is in the production of the materials and in the construction. Machines, on the other hand, are designed to convert energy efficiently – into motion, heat, messages and the like. Structures are the steady parts of the system, machines the dynamic. and the two are entirely interdependent. You cannot have an aeroplane without an airport, a generating turbine without foundations and a building, a chemical plant without a structure, cities without drains and power, even a plane without engines and a frame. But in detail structures and machines are often completely different.

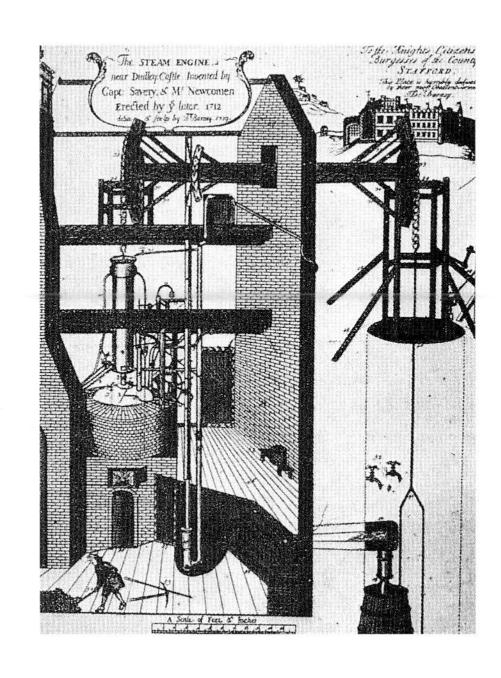

Opposite, top: Smeatonís classic study of water power; Bottom: Trevithick locomotive in Euston Square, the first steam carriage, 1808.

Above: Thomas Newcomen's atmospheric engine, the first engine run by steam power, 1712.

But does our society really understand and recognise the great contribution engineers have made to the modern world, and are still making today? Often, I think, the answer is 'No'.

Perhaps it is not surprising that the public, with its need to 'individualise' success, fixes its approbation on the package designer whose pencil co-ordinates and markets a design. In the design of machines, such as washing machines, computers and cars, the public often realises that the qualities which make the product excel are provided by engineers, but in my own field – the design of buildings – the engineer's role is less understood and therefore more undervalued. Nearly everyone in Britain, and perhaps elsewhere too, thinks that creative design in building is exclusively due to architects.

But in building the product is usually a complex one, requiring many skills in order to put many values into it. I am a building or structural engineer working in partnership with architects and others, each group bringing a body of knowledge, experience and sensibility to a common problem. Today construction is about big money and to handle that successfully calls for toughness and rigour. Autocracy or selfishness are not called for, but a system of collective decision making is essential. For such a partnership means mutual authority and shared recognition amongst the members of the building team.

Engineers, almost by their nature, excel at group work and avoid extravagant claims. They are very conscious that design usually requires many specialists who are designers in their own right and who put different qualities into the product. Engineers are sensitive of claiming sole authorship.

I am a structural engineer with an interest in building physics. Structural design is primarily concerned with the choice of form; the forces on that form and the analysis of its behaviour follow on from that choice. The whole process is influenced by the need for feasibility of execution, as success is proved by practicality.

Which qualities are essential to good structural engineering? Engineers should have an interest in the behaviour of materials and knowledge of the physics of the environment. We need to give value for money. As Herbert Hoover said: 'An engineer is a man who can do for one dollar what any fool can do for two.' And I think our ambition is to achieve elegance as well as value; 'elegance' in the mathematical sense, meaning economy as well as appropriateness. Appropriateness (or function) + economy = value. As a French aircraft designer once said, 'When you cannot remove any element then you have the right design.' And here of course we can learn from nature, in which structures have to be totally appropriate and mistakes become extinct.

The need for better living conditions for more people makes the engineer's struggle for efficiency worthwhile. I come from the generation which, largely due to accelerated technological changes in construction during the Second World War, came into engineering because of an interest in the efficient use of materials. This interest has pervaded my work as well as that of my colleagues and many others of my generation. We join a long line of engineers who have been working at long span, large space enclosures. The concrete shells of Nervi and Candela were modern versions of historic solutions; products of local materials and skills plus the advantages of mass in hot climates. Such structures are still relevant and we continue research into their construc-

Opposite: The world's first iron bridge at Calebrookdale, completed in 1779.

Top: Everyone knows Whistler's mother, but his Father, was certainly more productive. He came to Britain to learn from the Stephensons about railway engineering, returned to pioneer railways in America and went to Russia to start the construction of the Trans-Siberian railway.

tion methods. Buckminster Fuller emulated the Victorian engineers by following traditional forms, but he copied nature by reducing materials. The Pompidou Centre was not intended to be, but ended up as, a pastiche of the Crystal Palace – and what an expression of intermediate space that was. The energy input/output ratio of steel is high, so one becomes interested in timber. Thus the German inventor and designer, Frei Otto, using an equal mesh three-dimensional version of Robert Hooke's hanging chain model to define a tensile net form which, when reversed, provides a pure compression shell, then proposed that it be built with timber lathes. It would have collapsed. To act as a shell it needed diagonal cables to provide sheer stiffness but, because of its lightness, it was very economic to erect. It was test loaded with the town's dustbins and covered with a coated fabric.

To carry a force in tension is, of course, in material terms the most effective, and a very old, building solution. Reinforced with steel cable, such roofs can safely and economically cover several acres. The actual skin can be concrete - as in Calgary, designed by Jan Bobrowski with our partnership as proof engineers – or it can be PVC, PFTE-coated fibreglass, timber shingles, ceramic tiles or even, nowadays, stained glass. But most loading is from the wind and snow and thus cyclical. Steel cannot compare with the energy-storing characteristics of timber – but, if the steel wires are crimped to act in the more lightly loaded condition like a spring, it can store energy mechanically. We have used crimped steel mesh as a fabric for aviaries in Munich, San Diego and Hong Kong. Compare the wires' similarity to a spider's web.

But, to return again to timber, recent comparative studies have shown us that it is the strain energy characteristic which also reduces the cost so radically. Timber, in proportion to its weight, is comparable in strength and stiffness to high strength steels. But the problems with timber have always been in achieving an effective tension connection between members, and it is only since the discovery of epoxy adhesives, impregnated with fibres developed for windmill blades in the USA, that 'high strength' collinear connections have been made possible. We can now achieve 85 per cent full strength.

For many years carpenters have resisted removing any more of a tree than the bark. This is because the tree, subjected to sudden gusts of wind and not wishing its outer capillaries to buckle, orientates some of its fibres diagonally around the capillaries so that in containing the sap an element of longitudinal pre-stressing is given to the outer fibres (to the order of $14MN/m^2$), while the centre of the tree is in compression. When the wind blows, the outer capillaries stay in tension, which they are well able to carry. Combining advances in timber connections with the tensile properties of raw timber has made possible the hanging forms in the School for New Woodland Industries at Hooke Park.

Using a structure not only to resist forces but also actively to amend the internal environment becomes interesting. Light, of course, is the most powerful source of energy, and designing covered cities in the Arctic is simply an extreme version of this. In order to achieve a satisfactory all-year-round environment under such a cover, the quality of light is all important and the quality is dependent on as much of the spectrum as possible being transmitted through it. This is why glass is used for windows, even though it does reduce the ultraviolet part of the spectrum, thereby creating a 'greenhouse' effect.

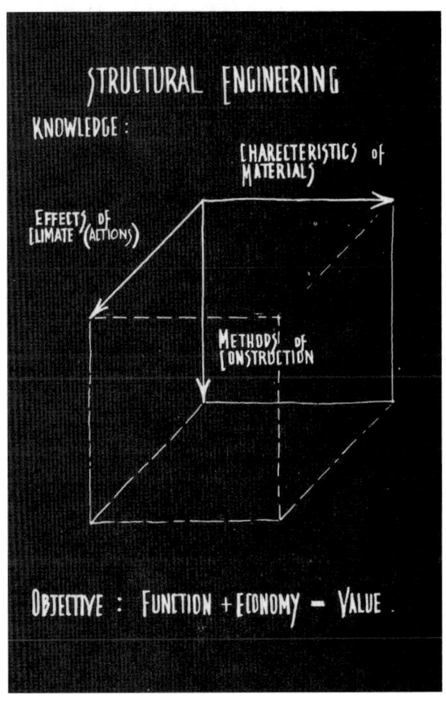

Opposite, top: SS Great Britain, engineer I K Brunel, c1843; Bottom: Thomas Telford's bridge over the Menai Straits, 1818–26.

Above, top: the raising of the transept ribs of the Crystal Palace, Hyde Park (Paxton/Fox Henderson) 1850. Bottom: meeting the Happold criteria – Function + Economy = Value.

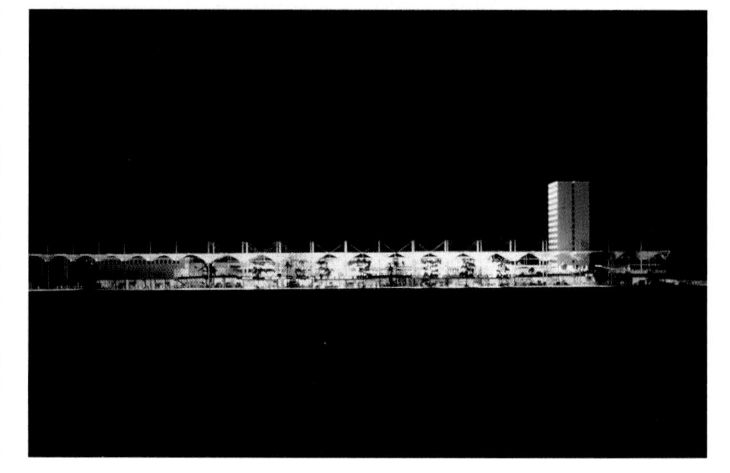

Studies have been carried out for an air-supported cover over a proposed 36-acre city in the Athabasca region of Alberta. These show how some of the new laminates transmit more of the light spectrum than glass – making possible even the growing of grass in the Arctic. Alas, the collapse of the oil market caused the project to be dropped. This development has moved slowly since then, but one of the German manufacturers has used this laminate as a cover for a conservatory for tropical birds and alligators at Arnhem in Holland, and we used the lighting aspects of the concept for a roof in Brunei. It seems to perform very well.

Of course, our bubble idea is still relatively crude and needs further work. But one cannot help but be interested in further developing the possibilities of utilising energy from light. After all, this is only the old idea of the conservatory or greenhouse. When it is too hot we shade ourselves, like plants do with their leaves. But we still cannot build firm organic substances from carbon dioxide, water and light, as plants do. We must find out how to do this, before the slight surplus store of carbon – coal, oil, gas and timber – which has been built up over millions of years and which at the moment provides most of our structural materials, is finally depleted.

Engineers are moving towards the solution. In fact organic materials are becoming more and more important. And just as 200 years ago those who resisted change thought iron and steel 'inhuman' materials, today people say the same about plastics. Yet such materials, in composites and laminates, are essential materials in aerospace and their use is growing.

All this is about change. Yet much engineering is at a simple level. The pace of advance in technology is generally set not by the most brilliant and able engineers but by the capacity of the individual – engineer or skilled mechanic – to master and use an improvement efficiently and harmoniously. My last example illustrates this: an exposed bridge over a busy road in a town. A simple steel trussed solution can be prefabricated and erected in a matter of hours. Then it can be clad like the building to reduce exposure and deflect the wind. A simple piece of street furniture.

You may have noticed that most of the ideas in these projects were first developed for overseas clients. Why is new technology not better utilised in Britain? The Prince of Wales is promoting community architecture in the cities. Yet there are few problems of the inner cities which could not be solved by the economic well-being which productive industry could provide. It is engineers who could bring about this. Our car industry is a failure, yet we have engineers who design the fastest, safest and most reliable cars in the world. If only these two groups could work together in Britain as equals, they could succeed.

Yet our intellectual and economic class systems are such that this seems impossible. We have the skills but not the business or social structure to enable our engineering expertise to revitalise our industry. Perhaps the failure is managerial. What is common to most successful overseas technical enterprises is the inevitability of collective decision making. Could it be that our national belief that someone should be in charge – preferably on an annual financial basis – inhibits this? Maybe those Quaker industrialists with their collective long-term ambitions were right and we should consider returning to those lost managerial values.

Opposite, clockwise from top left: Palm House at Kew, c1844 (Richard Turner, engineer; Decimus Burton, architect); Diplomatic Club, Riyadh (Omrania, Frei Otto, Buro Happold); Buckminster Fuller's, American Pavilion, Montreal, c1967; Britannia Bridge under construction c1847 (Robert Stephenson, engineer); Basildon Town Centre (Michael Hopkins, Buro Happold); San Diego Aviary (Buss Silvers with Jorg Gribl, Buro Happold); Mannheim, Bundesgartenschau (Carlfried Mutschler and Partners, Frei Otto, Ove Arup & Partners); Saddledome, Calgary (Jan Bobrowski).

Above, top: Proof test at La Jacaranda nightclub, Mexico, 1957 (Felix Candela, engineer); Bottom: Pier Luigi Nervi's, the Palace of Sport, Rome, 1956–7.

REMOVALS
01-603-6611

Work at Ove Arup & Partners

Sir Basil Spence

Life is often touched by the combination of luck and location, and certainly Ted Happold was doubly fortunate to meet Basil Spence in his role as Hoffman Professor at the University of Leeds when Ted was studying in the Civil Engineering Department. He was advised by Spence, who had befriended him, to join the already legendary practice of Ove Arup and Partners, and under Povl Ahm, who enjoyed the direct influence of Ove Arup himself, he worked on the innovative structure for Coventry Cathedral. Spence also encouraged Happold to visit Finland to acquaint himself with the work of Alvar Aalto. Prior to joining Arup's office as a student, Happold presented himself to Alvar Aalto, and brief stay at his summer house and studio certainly influenced his subsequent approach to the integration of building engineering and architecture.

In 1961, after two years in New York, Ted returned to England to Arup's where he was redeployed on the demanding programme of work in Spence's considerable portfolio. This included a number of university projects, particularly the Science Building at Exeter, and, later, Structures 3's Kensington Town Hall which was the culmination of Spence's Knightsbridge style. However, the two buildings that Ted was most involved with were the British Embassy in Rome and Knightsbridge Cavalry Barracks in Hyde Park.

The initial brief for the barracks required, within stringent scales of expenditure, quarters for some 500 officers and men, with associated messing; recreational areas; stores; offices; workshop, educational and other facilities; together with a riding school and stabling for some 270 horses. The complex has not been without its share of critics. The tower specifically, and the density of development as a whole, concerned Londoners as it created with the Hilton and the Royal Lancaster, 'a trio of interlopers' in the park peninsular, becalmed in a nineteenth-century urban aesthetic.

The architectural vernacular followed very much the aesthetic used by Spence at Sussex University. Indeed all the buildings other than the tower, each of different function and scale are unified by a common structural system comprising an in situ reinforced concrete frame carried on piled or strip foundations. This is partly clad with very precise brickwork which, where exposed, is fair faced both inside and out. Further unity is given to the tower by the use of cylindrical concrete shell units. The shell hangs off secondary beams which bear on the main beams. The hanger beams were pre-cast first and then cast into the shells as they were paired. It is interesting that Povl Ahm, writing in the *Arup Journal* in March 1966, puts his finger on the problem of construction and form for both Sussex University and the Knightsbridge Barracks. 'Sir Basil Spence has attempted to use the shell forms aesthetically, but economic and other considerations prevented Arup's from following him in actual structural form. The result was a prefabricated solution of beams and curved slabs. The question therefore is whether it is justifiable to use form for its own sake ... if it creates the desired architectural effect!'

Opposite: Work with Ove Arup & Partners for Sir Basil Spence.

Clockwise from top left: The interior of Coventry Cathedral, looking towards the sanctuary with the Sutherland tapestry; The Hyde Park Cavalry Barracks from the Monpeliers; The Cavalry Barracks interior stabling area; View of the barracks looking towards Knightsbridge; Exterior of the British Embassy in Rome.

Above, top: Ted's modest beginnings at Arups involved working for Povl Ahm on the John Hutton screen at Coventry Cathedral; Bottom: Later work with Spence included a number of university projects, including the Science Building at the University of Exeter.

British Embassy, Rome

Any description of the architecture of the Embassy cannot be divorced from a description of its location – on the Via Venti Settembre (previously the Via Pia) and alongside the Aurelian wall. The jewel in the crown on this famous street is Michelangelo's gate, the Porta Pia, which is built into the Aurelian wall. The main Embassy building is the Chancery, a two-storey square ring structure on cruciform pilotti with an entrance and lift core area carried down to ground-floor level. All services are housed in a separate building by the Aurelian wall. Roads, car parks, boundary walls and three external pools complete the works. The concrete frame consists of fifteen cruciform-shaped columns from the top, of which the first-floor octagonal cell structures cantilever. The cell structures have fair-faced concrete cruciform beams set over the heads of the columns with a double-skin floor system enclosing the services. A central reinforced spine beam runs through the building and each bay is separated from the next by a transverse concrete wall acting as a room divider. The second-floor slab cantilevers over the first floor and the perimeter walls support travertine panels clear of the structure.

The project was particularly instructive for Happold as for the first time he had to grapple with the contractual and constructional problems of a high-quality building outside Britain.

Left: Kensington Town Hall. From top: Interior view of the main hall; The structure of the undercroft (the buildings largely detailed in Spence's Knightsbridge style); Red brick cladding on a reinforced concrete frame.

Opposite, right from top: photographs showing model of the British Embassy in Rome, with the Porta Pia; Middle right: Courtyard view; Bottom: The ceremonial access stairs from the embassy court.

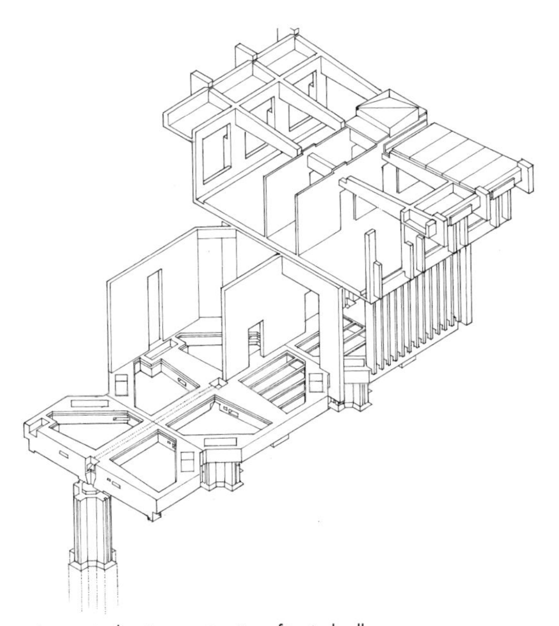

Isometric showing construction of typical cells.

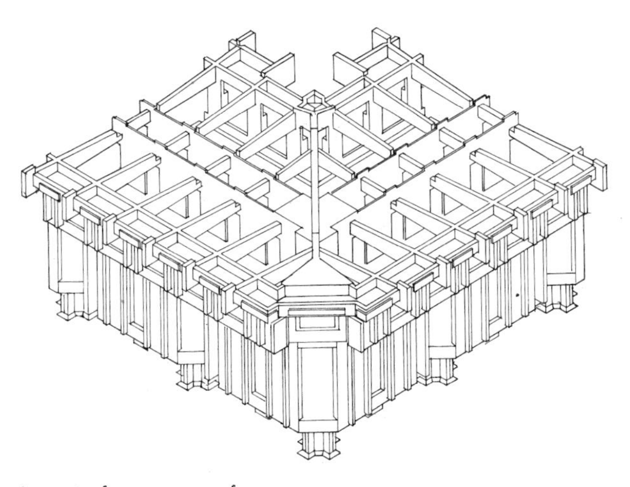

Isometric of corner up to roof

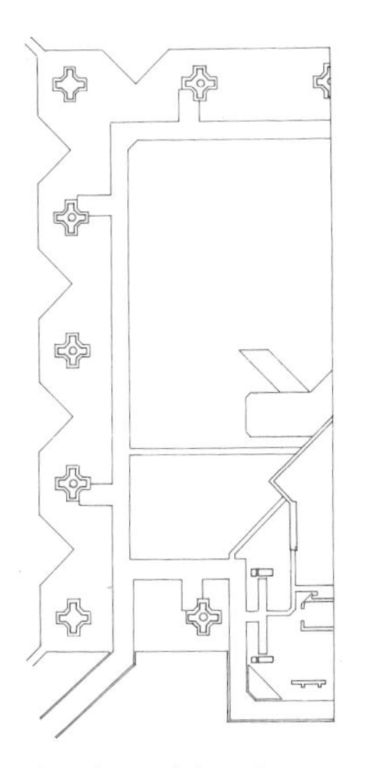

Plan of Ground/First Floor.

Ted Hollamby

When Ted Hollamby turned down the chance to take over as Chief Architect at Livingston New Town and settled for keeping his London roots and taking up the post of Chief Architect for Lambeth, he inadvertently enriched the scope for a socially motivated engineer in the Arup organisation, Ted Happold. Hollamby was a graduate of Sir Leslie Martin's time at the GLC, a period when many of the most talented young architects returned from the war intent on providing post-war Britain with a socialist Utopia. Many moved on to distinguished private practice; some, like Hollamby, stayed the course in the public sector, politically and socially motivated to a degree that is now only a fond memory in post-Thatcherite Britain.

Make no mistake, Hollamby was no local authority cipher but a man of conviction, sensitivity and no little skill. His calibre and historic roots were underlined by his loving restoration of the 'Red House' (built by Philip Webb for William Morris), where he has lived since 1953.

When Hollamby took over at Lambeth his strong belief in the necessity of a settled team, hand-picked for continuity and consistency, led him to an old friend at Arup's, Peter Dunican. His requirement was for a bright, socially motivated engineer who would spend a great deal of time in his department setting up procedures for a joint and continuous working relationship with the Arup organisation. He had been inspired as a young architect by the brilliance of Felix Samuely and was looking for a collaborator of similar potential. Dunican's recommendation was inspired. Ted Happold had the credentials for the assignment, plus he had strong socialist and Quaker principles and was from a family equally motivated and experienced in the field of public housing.

The two Teds became firm friends and Lambeth's dynamic programme honed one of Happold's strengths – his ability to inspire and create a family or families of engineers dedicated to the pursuit of better design. Three of his original partners at Buro Happold – John Morrison, Peter Buckthorpe and Rod Macdonald – cut their teeth on the stepped housing at Central Hill. Predictably Ted's first assignment was for a children's lavatory in Vauxhall Park (looking rather like a float from the Mardi Gras), but perhaps more typical of the variety of projects processed by Arup's in this period are West Norwood Library, a mixed-development in Kennington, Central Hill housing and an extraordinary subterranean boiler house serving Myatt's Field.

What Ted Hollamby's work at Lambeth required from a structural engineer was the knowledge and experience to handle, in an elegant and integrated way, the many conventional elements that made up his buildings, slabs, beams and trusses in timber, steel, or concrete – and to do so within the stringent budgets implicit in public sector projects. Ted's ingenuity and imagination contributed greatly to their joint sucess and out of this close working relationship grew a friendship between two people who enjoyed the same philosophy and showed a similar taste for the gothic and a great sense of fun. Their trip to the Swiss National Exhibition in Lausanne in 1964 was significant in crystallising Happold's interest in long span roofs – the hanging plywood roofs he saw there provided the inspiration for the stainless steel roofs he proposed for the City Club at Milton Keynes – and the sight of Frei Otto's tent structures in the exhibition (made in collaboration with Peter Stromeyer) provided Happold with the initial inspiration for nearly twenty years of engineering, research and development.

Opposite, clockwise from top: View of Central Hill housing, Lambeth; Architectural model of the Central Hill project; underground boiler house, Myatt's Field; The interior and the structure above ground.

Above, from top: The children's lavatory at Vauxhall Gardens; Multi-use development, Kennington, facing Harwood Park and the Imperial War Museum.

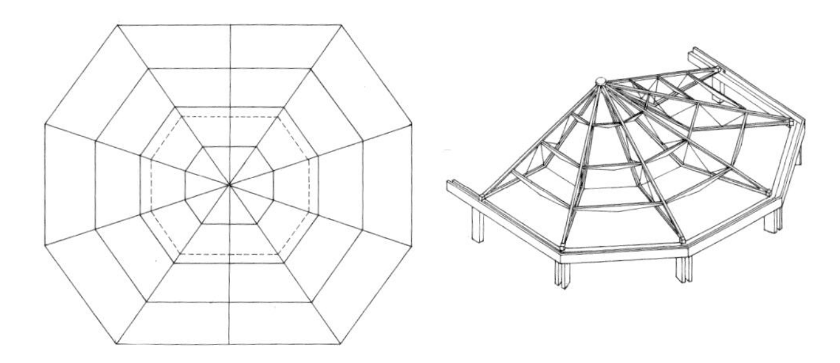

Triangulated steel trusses of rectangular hollow section with fish bellied perlins

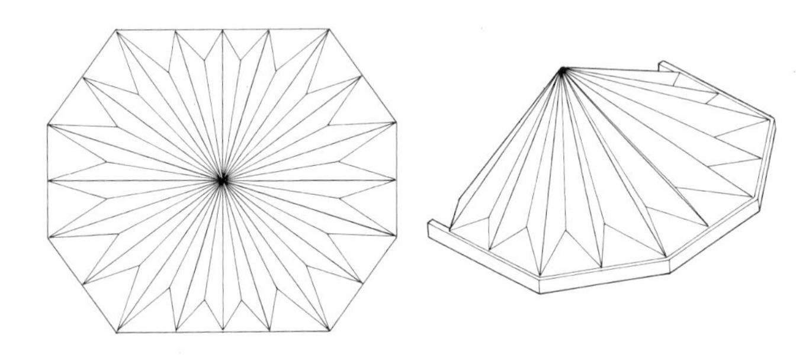

The second scheme used stressed skin plywood folded platis with a concrete ring beam supported on conctete columns

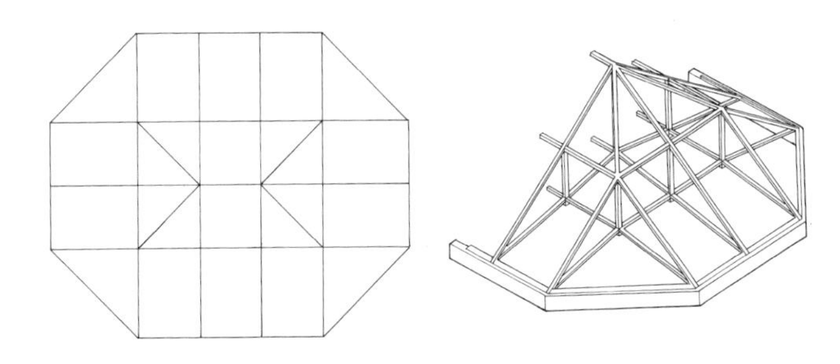

A pitched roof with two way spanning timber trusses: ring beam acting as gutter

1. West Norwood Library on the selected configuration. Triangular Steel Trusses of rectangular hollow section with fish bellied purlins were used a separate stage lighting support structure was developed in view of the large size of ducting required twin columns were used with duct space between Forces from the cantilever projection room were taken in the ring beam and column structures.

2. This rejected scheme used stressed slein plywood folded plates. The planning requirement was such that the octagon had to be stretched from the square shape originally selected.

3. The penultimate structural option the services were to be included in the roof space and the stage lighting was to be hung from the roof. Rainwater gutters were introduced and as the outer covering of the roof was copper the roof was designed to minimise surface area.

West Norwood Library and Community Hall

The library, planned as a single-storey structure surrounding a courtyard, shares its entrance with the Community Hall. The three-storey-high hall structure shares a stretched octagonal form with the bar and restaurant below. The library is constructed of load-bearing brickwork with square, steel-box-section columns forming the structure to the sliding, glazed walls of the courtyard. The mono-pitched roof has timber joists on steel beams. The Commuity Hall is a reinforced concrete structure with brick infill walls. The roof was originally designed as a stressed skin plywood pyramid, each face formed as a tetrahedron, with three faces inside and one externally. The varying options are shown left and the final solution was built with triangular steel hip trusses. All roofs are covered in copper.

Kennington Mixed-use Development

The development consists of apartments, housing above shops, the office of the Registrar of Births and Deaths, a doctors' group practice, a chemists, and a chiropody clinic. It is situated on a prominent but near-impossible half-acre corner site facing Harwood Park and the Imperial War Museum. The complex structure consists of a string of linked towers of varying height in a serpentine form built in reinforced concrete with a textured finish and lightweight infill panels. Heating and hot water services are fed from the boiler house of the adjoining public baths.

Above: Structural options for the roof at West Norwood Library.

Opposite, clockwise from top left: Circulation West Norwood Library; The Community Hall and Projection Room; View of the library from the seating area; Kennington mixed-use development; site plan showing the Kennington development – apartments, shops, registrar, group practice, chiropody clinic; roof configuration, Community Hall, West Norwood.

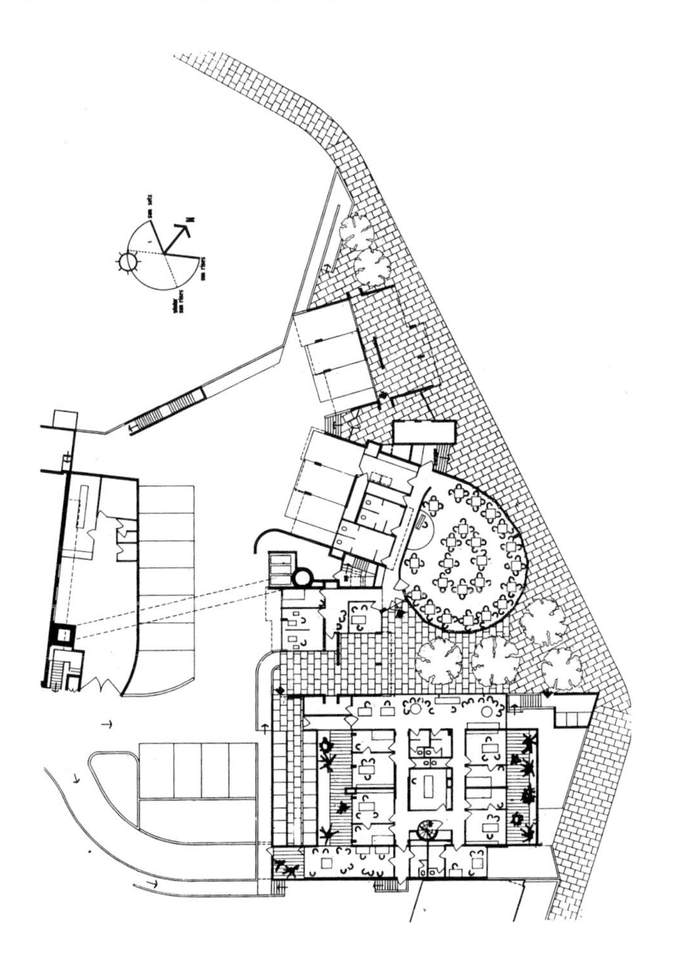

Renton Howard Wood Levin

One of the projects completed during Happold's period at Arups that particularly captured his interest was a Taylor Woodrow Property Company development, won in competition with Renton Howard Wood as architects and Ove Arup as engineers. It involved the redevelopment of St Katherine's Dock, adjacent to the Tower of London, after its closure in 1968. The warehouses (designed by Hardwick in 1819 and completed by Aitchieson) and the dock itself (designed by Telford) were internationally recognised. Their redevelopment was sympathetically handled, retaining warehouse I intact, modifying warehouse B and dredging the water area to form a marina. The whole assembly offered a superb example of nineteenth-century engineering and its lessons were not lost on Happold. It most certainly inspired his continuing interest in the appraisal of existing structures. He organised a conference at the University of Bath in 1984 on structural appraisal and the refurbishment of existing structures as part of his duty in chairing the Institution of Structural Engineers' Task Group. *Appraisal of Existing Structures* published initially in 1988, then substantially revised and re-issued just before his death.

Renton Howard Wood and Levin were also pioneers, with Peter Moro and Chamberlain Powell and Bon, of new theatre design in post-war Britain. Their work on the Crucible in Sheffield (with its full thrust main auditorium and studio theatre) and the new arts theatre on the Warwick University campus were experiments in new theatre forms, developing ideas that were slowly emerging as templates for the auditoria of the National Theatre. The Warwick project brought Ted into contact for the first time with John Bury, at that time the chief designer for the RSC, whose work at the Barbican and the National Theatre reinforced his reputation in the dual role of Britain's pre-eminent set designer and theatre consultant.

Opposite, above: The Crucible Theatre Sheffield showing its full thrust auditorium. Below: the Crucible foyer and Circulation space. Above, clockwise from the top: Minimalist steel frame on an hexagonal grid for Warwick University Arts centre. Detail of Cast Iron columns St Katherines Dock, London Warehouse buildings. Model of the Re-development showing water area adapted as a marina view of I warehouse with its Italianesque campanile by Aitcheson.

Trevor Dannatt

Ted's relationship with Trevor Dannatt was as significant as it was long standing. The first project with Dannatt came about in a curious way, as Happold recalled in *Patterns* (Buro Happold's in-house journal):

'I went to a Quaker school in York called Bootham and when in the early sixties the name appeared on the job list in Ove Arup and Partners I asked if I could be the engineer for it. The architect was Trevor Dannatt, and a fellow engineer, John Martin, not only looked after Trevor's work but had digs in his house. John is a generous man, he let me do it and a very close friendship between Trevor and myself developed. We worked together not only on Bootham but also on the Friends Meeting House at Blackheath. So when in 1966 the government of Saudi Arabia held a limited competition for a conference centre and government guest house and Trevor was asked to enter, I went along as the engineer doing the work at the weekends.'

When Ted wrote about Bootham School in the *Arup Journal* in March 1966 he said:

'I think the most satisfying building that our group at Arup's worked on in 1964 was Bootham School. There are two reasons for this feeling. Firstly, my family have many connections with the school and I was a pupil there myself and, secondly, I found a great understanding and sympathy with what the architect was trying to achieve and by and large I think he has achieved it. The roof was really the first part of the building designed. Trevor Dannatt felt it was important that in order to achieve a uni-directional feeling in the hall related to the shape and a centralised feeling related to worship he needed to form a higher central roof ringed with clerestory lighting. He also felt strongly that the hall itself should have no columns within the perimeter.'

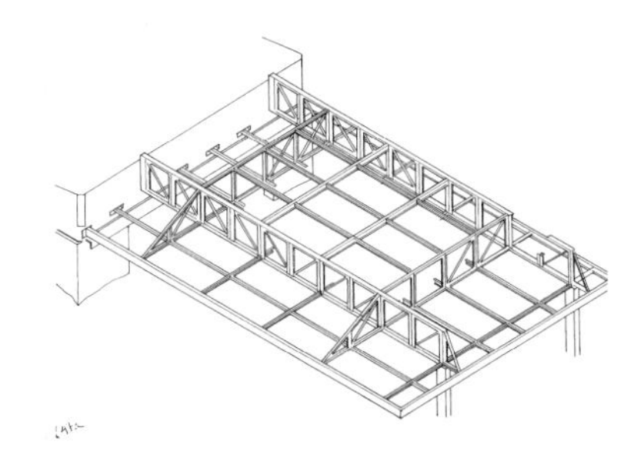

Bootham School Assembly Hall: roof structure

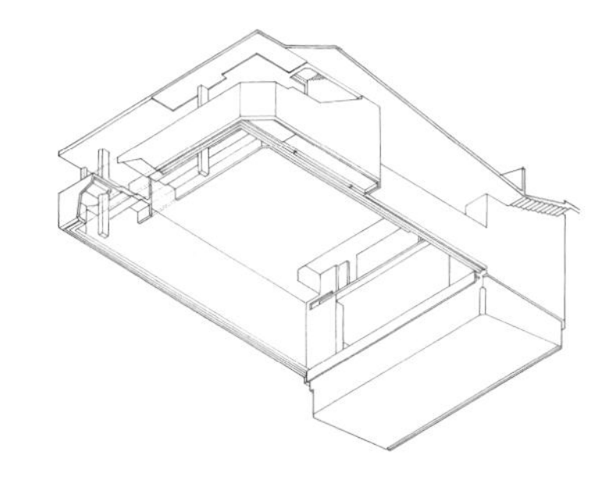

Bootham Assembly Hall: main structure

The Conference Centre in Riyadh followed in 1967 and is described as part of Happold's journey through the Islamic world later in this narrative. The project was a triumph, both technically and in overcoming the difficulties that always arise when working in an unknown culture and construction environment. This project was started at a time when very few Europeans were working in Saudi Arabia and perhaps here it is appropriate to mention Arups resident engineer from Structures 3, Peter Woodward, a close friend and colleague of Ted Happold whose personal tenacity achieved a standard of workmanship that was quite extraordinary in the difficult conditions that prevailed at that time in Saudi Arabia.

In 1970–2 the Quaker meeting house in Blackheath was completed; some years later work on the British Embassy and housing in the Diplomatic Quarter of Riyadh cemented almost two decades of collaboration. Their work in the Middle East allowed both parties to contribute to a research programme concerned with materials, technology and climate – a body of knowledge which increased in sophistication, maturity and understanding. This becomes very apparent when one compares the earlier work in Riyadh with the more recently completed Embassy compound.

Opposite, from top: Assembly Hall at Bootham School, York, 1964, with the Minster in the background; 1967 entry for a Cor-ten steel bridge sponsored by US Steel Bridge 1967 and (designed with Trevor Dannatt, Ted Happold and Ole Vanggaard); The Intercontinental Hotel and Conference Centre, Riyadh 1967.

Above, from top: Trevor Dannatt and Ted Happold in Beirut in 1967; Axonometric of the roof structure and axonometric of the main structure, Bootham School Assembly Hall, York.

View of Conference Centre and Intercontinental Hotel, Riyadh.

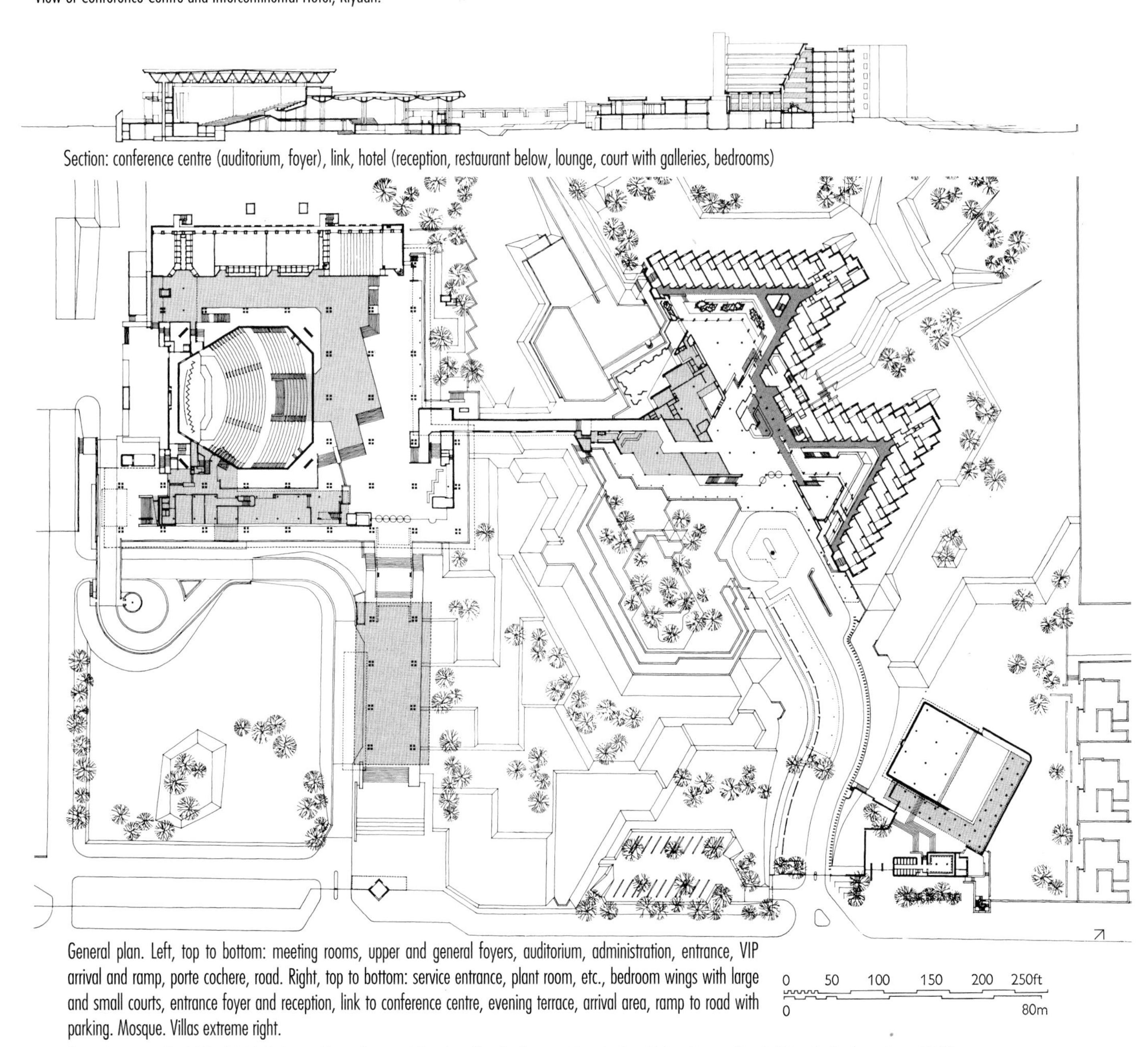

Section: conference centre (auditorium, foyer), link, hotel (reception, restaurant below, lounge, court with galleries, bedrooms)

General plan. Left, top to bottom: meeting rooms, upper and general foyers, auditorium, administration, entrance, VIP arrival and ramp, porte cochere, road. Right, top to bottom: service entrance, plant room, etc., bedroom wings with large and small courts, entrance foyer and reception, link to conference centre, evening terrace, arrival area, ramp to road with parking. Mosque. Villas extreme right.

Opposite, top and middle: British Embassy Chancellery and Housing, Riyadh; Bottom: Victoria Gate Visitor Centre, Royal Britannia Gardens, Kew, 1992.

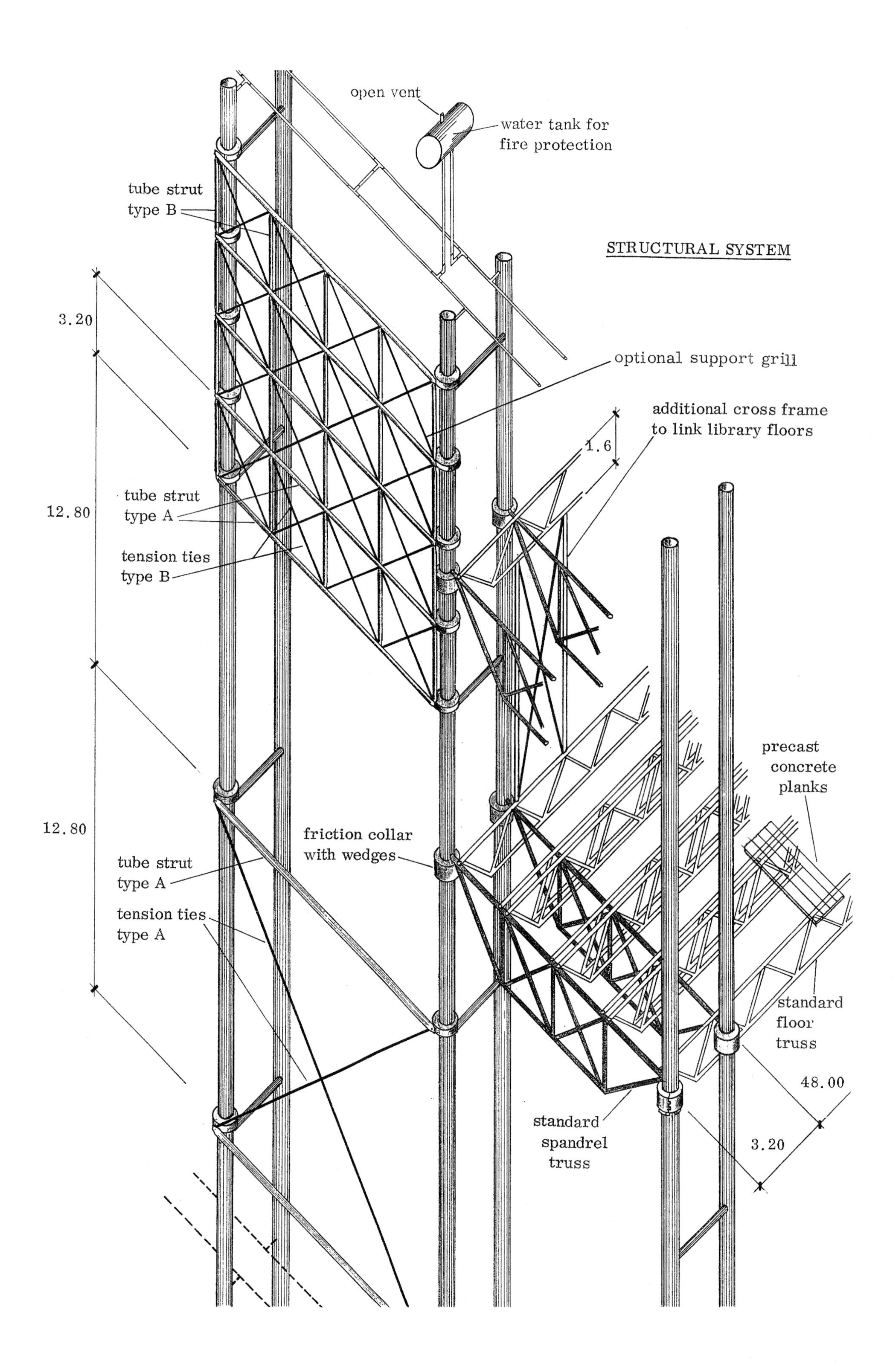
open vent
water tank for fire protection
STRUCTURAL SYSTEM
tube strut type B
3.20
optional support grill
additional cross frame to link library floors
1.6
tube strut type A
12.80
tension ties type B
precast concrete planks
12.80
friction collar with wedges
tube strut type A
tension ties type A
standard floor truss
48.00
standard spandrel truss
3.20

Beaubourg

As a single building, Beaubourg, perhaps more than any other, demonstrates the outstanding qualities Ted Happold injected into the building process: perception; conceptual, engineering and negotiating skills; and design leadership.

Ted instigated the competition entry after an abortive sortie with Richard and Su Rogers in the competition for the Stamford Bridge stand which, in Richard Rogers' view, was an all-round catastrophe in that neither client nor consultants were ever convincingly on the same wavelength. He was touched and surprised when Ted came back a little later with a proposition to enter the open competition for Beaubourg. At first Arup's were not going to pursue the competition and Richard Rogers was also ambivalent about buying a ticket for the lottery, but Su and Renzo Piano liked the idea and Richard concurred. The competition entry was late to gestate but generated around a clear image the team wanted to explore – an idea stemming from the Joan Littlewood/Cedric Price Fun Palace and the freewheeling sixties culture of Archigram. The solution was brilliantly graphic in combining the principles of loose-fit technology and engineering versatility, demonstrated by the movable floors and the articulation of a pedestrian transportation system designed to become the greatest free ride in Europe. In essence, the competition entry, much of it conceived around Ted's kitchen table at Gloucester Crescent, was presented as a gigantic interactive information machine.

The impossible happened – out of 687 entries the fledgling practice of Piano Rogers and the equally youthful members of Structures 3 (Ted Happold, Lennart Grut, Peter Rice and Michael Sergeant) won the day. Fortunately, the messianic zeal and contracts experience of Ted Happold and the gravitas of Arups kept the project on track despite a difficult two years during which Ted mobilised the team, set up an obligatory Paris office and carried out a complex contract negotiation that had to recognise the impossibility of separating execution from intention – a process then unfamiliar in France.

Richard Rogers is in no doubt that Happold's efforts on behalf of both architects and engineers – and, his strong personal relationship with Robert Bordaz, the President of the client body – were pivotal in effectively clearing the way for the teams to operate properly and effectively on the project. Bordaz remarked that 'the brilliance of the competition entry was one thing, but it was the toughness of the negotiator that ensured the project was built. Ted's last act in Paris ensuring the project's implementation was to oversee the foundation contract and the excavation of the largest hole in Paris.

During this hectic two-year period of constant commuting, Ted's energy and tenacity in extracting the best possible financial package for the consultants ruffled a number of Gallic feathers and the Arup hierarchy decided to try a fresh scenario. Ted was asked to return full time to Structures 3 in London. He was desolated to relinquish day-to-day control to the Peter Rice/Graham Wood team in Paris and bitterly disappointed that his contribution to the project was not publicly recognised.

It was not until 1995, in his very generous review in the *RSA Journal* of Peter Rice's book *An Engineer Imagines* that Ted revealed publicly the mixture of

Opposite: Ted Happold's competition drawing for the Beaubourg competition.

Above, top: Three of the Arup team for the Beaubourg competition – Lennart Grut, Ted Happold and Peter Rice; Competition model; the Cedric Price/Joan Littlewood concept for a Fun Palace;

Page 52, clockwise from top left: Two shots of the main structure during construction; A graphic illustration of the building's popularity; Two shots showing the casting of the gerberettes at Krupps.

Page 53, clockwise from top left: The final model; A view from the piazza (Breughel revisited); The escalator on the front façade; Typical corner adjacent to the escape staircases; View through the gerberettes; A view of Paris from the Sculpture Gallery.

elation and anguish (in equal measure) that Beaubourg had generated over time on his psyche. He sought in that review to correct elements of misinformation that always seem to bedevil high-profile projects. The three elements that dominated the structural concept for Beaubourg are vividly expressed in Happold's competition drawings: an exposed steel scaffolding which he likened to a Victorian structure, movable floors which would have animated the interior of the museum immeasurably if they had been implemented, and the use of castings for the joints of the structure. The latter did not emanate from Rice's trip to Osaka but were the product of Ted's discussions with Kenzo Tange's partner, Koji Kameya, two years earlier as they waited in Kuwait to sign the contract for the Sports Centre. Ted's commitment to Beaubourg was immense, from conception to completion. There can be no argument that history will certainly document that without Happold Beaubourg as we know it would not have happened and Paris would be poorer for losing this single physical manifestation of the sixties' preoccupation with loose-fit technology and the popularisation of cultural access.

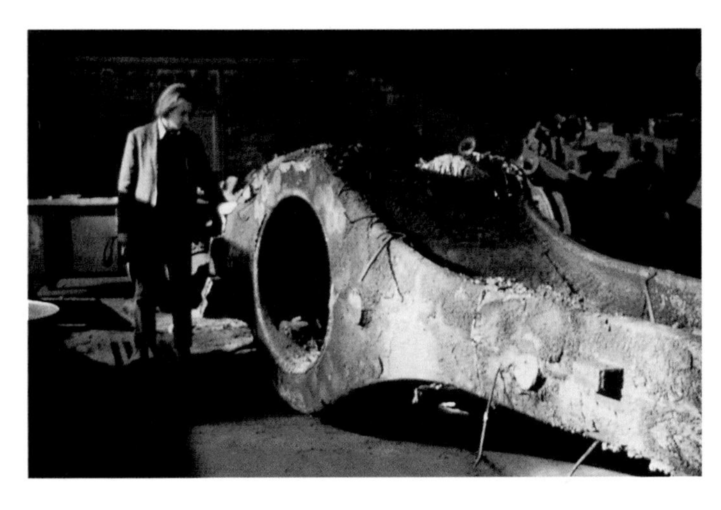

Lightweight Structures

The engineering theme which engaged Ted Happold's enthusiasm longest was probably that of lightweight structures. In one sense he shared this with most structural engineers since producing the lightest structure to fulfil a given function is always a limit towards which the design engineer can aspire. This is evident in many long span roofs but most of all, perhaps, in bridges where the goal becomes the longest span possible. The limits in both cases are set by the strength of the material used and the structural efficiency of the way in which is used. Since tension is the most efficient way of exploiting the strength of a given material, it is no surprise that the longest bridge spans have usually been on suspension bridges. The suspension system, however, has the disadvantage of being very flexible in the direction perpendicular to the line of the cable or chain and this has generally made it unsuitable for long spans in buildings.

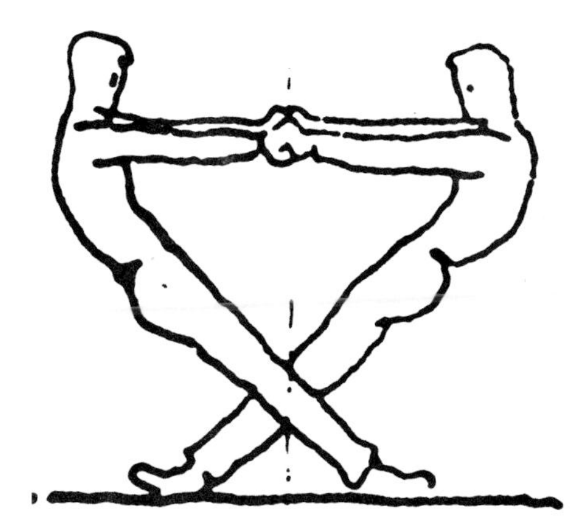

This situation prevailed until 1952 when the structural engineer Severud and architect M Nowicki developed for the Raleigh Arena a workable means of providing out-of-plane stiffness to a grid of steel cables. One set of parallel, hanging cables was stabilised by an orthogonal set pulled taut to create a saddle surface. While the Raleigh Arena was hardly the usual building type to be given architectural accolades – it is the livestock arena for North Carolina – it quickly gained acclaim from many engineers and architects. Eero Saarinen collaborated with Severud to create the roofs over the Yale Ice Hockey Rink and the Dulles Airport Terminal (both completed in 1958), using concrete to stiffen and stabilise the single-curvature steel suspension cables.

Severud's pioneering work with cable structures influenced Ted Happold when he was looking for further work experience in the United States, and in April 1959 he began an eighteen-month period working in the New York office of Severud, Elsted and Krueger. Although he did not work on any cable structures there, he was influenced by Severud's attitude towards engineering problems – on the one hand, the boldness to dare to tackle the unprecedented and, on the other, the need to base innovative solutions on sound engineering principles.

Ted returned to Arups in 1961 when their work on the Sydney Opera House was consolidating their reputation for helping innovative architects with unusual structures. For Ted, however, it was not until 1967 that he entered full bloodedly into the world of tension structures. His trip at that time to Riyadh with Trevor Dannatt prefaced the most significant meeting of his professional life. He had seen and admired Frei Otto's work in the 1964 Exposition in Lausanne, which he had visited with Ted Hollamby. He had also glimpsed manifestations of the formidable Gutbrod/Otto collaboration in the German Pavilion at the Montreal Exposition (designed with engineers Leonhardt and Andra) – together with the reputation of Arups. In Riyadh, Ted's entrepreneurial personality and his apparent command of the situation – led Gutbrod and Otto to entrust him with the engineering for their new project in Mecca.

Opposite, clockwise from top left: Cable net roofs covering a restaurant at the Lausanne Exhibition 1964 (Frei Otto); The German Pavilion, Montreal 1967 (Rolf Gutbrod, with Frei Otto); Two shots of the Munich Olympic Stadium 1972 (Günter Behnisch with Frei Otto); Two views of the Institute of Lightweight Structures on the campus of Stuttgart University at Vaihingen – erection of the net and completed structure, 1965.

Above: Severud's sketch for Raleigh, explaining the structural action; Saddle-shape roof of the Raleigh Arena by Severud and Nowicki 1952; The Ice Hockey Rink at Yale, a view of the interior, Eero Saarinen and Severud, 1958.

Frei Otto had also been inspired by Severud and Nowicki's Raleigh Arena as a young researcher at Stuttgart University, and his fascination with the forms

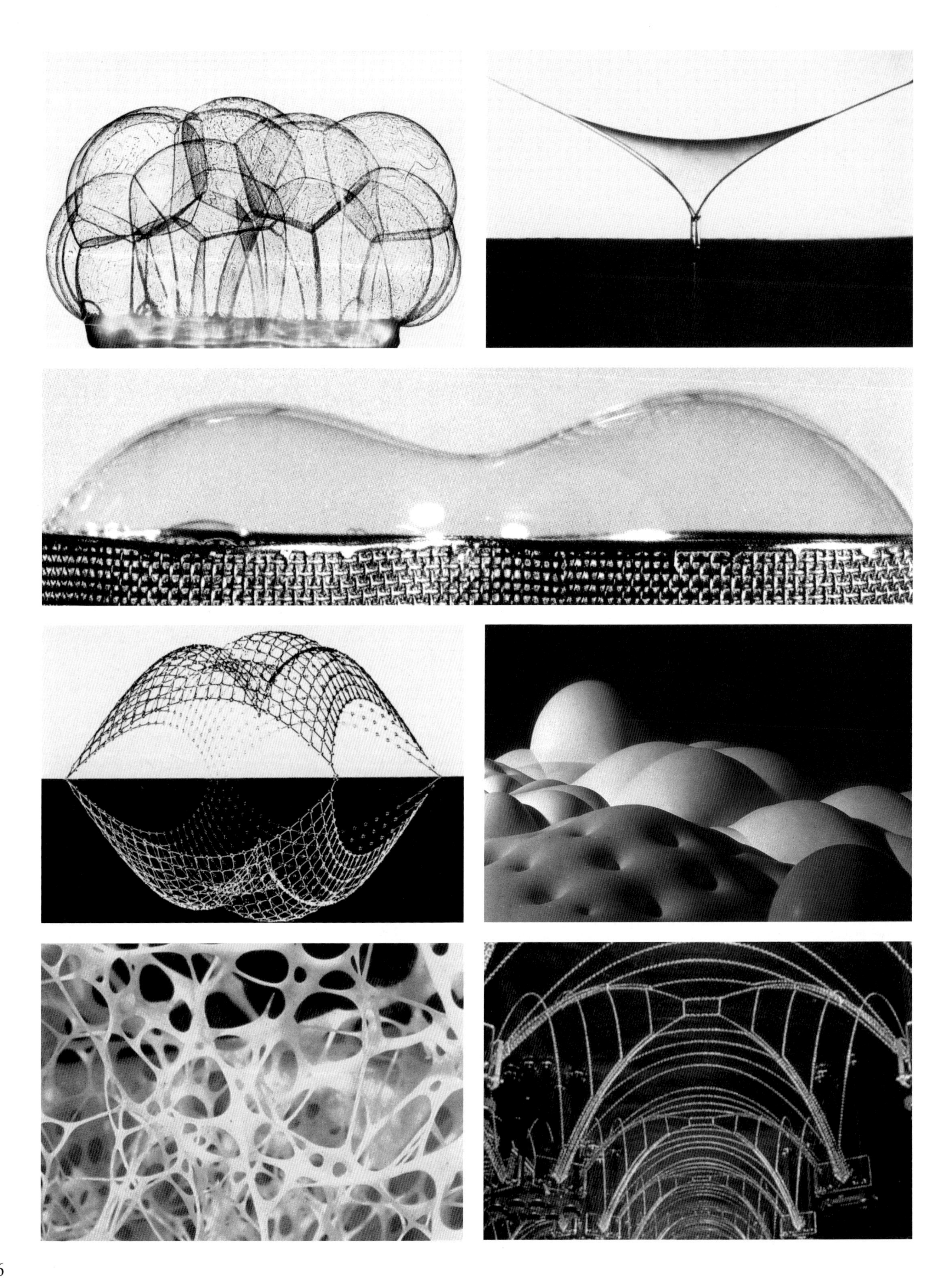

of structures that find their own shape became the focus of his doctoral work, culminating in the publication *Das hängende Dach* (The Hanging Roof) in 1959. Otto made several small, experimental tensile roofs and canopies but the main challenge at that time was how one would set about designing a large tensile surface structure. Conventional structural calculations were too complex to solve in a reasonable time, principally because the geometry of their surfaces could not be defined using convenient equations and, being statically indeterminate, information obtained from small models was virtually impossible to scale up to full size. Another difficulty was the low strength and durability of woven fabrics in the 1960s.

Otto's breakthrough was to use cable nets. These separated the load-bearing and the envelope functions of the roof by using a mesh of steel cables to support a woven polyester fabric which carried no significant stresses. By using an orthogonal flat mesh of cables, surface forms of quite considerable complexity could be arrived at on the basis of simple 'soap bubble' saddle forms with minimal surfaces. Minimal surfaces have equal and opposite principal radii of curvature which are orthogonal and so practical load-bearing structures can be achieved by equal mesh cable nets formed as a series of saddle surfaces. This allows the resulting geometry of the surface to be of much greater complexity in three dimensions than surfaces determined by more usual mathematical equations. After Lausanne in 1964 Otto worked on a family of structures: firstly, a prototype for his newly formed Institut für Leichte Flächentragwerke (Institute for Lightweight Structures) in Stuttgart in which he developed design procedures for the remarkable German Pavilion at Expo '67 in Montreal – and, finally, the roofs for the stadia at the 1972 Munich Olympics, working this time with Günter Behnisch, also engineered by Leonhardt and Andra).

During the Mecca project, and for twenty-five years thereafter, Ted was a regular visitor to Otto's Institute in Stuttgart. He was captivated by the utterly experimental approach to developing structural form – something extremely rare in Britain – and the perseverance shown by Otto in transforming an almost childlike passion for model-making into one of the major design revolutions that the history of structural engineering has seen. Otto had overcome the main impasse that held back the development of tension structures in the 1950s. He perfected a large number of experimental techniques for creating the shapes of different types of stressed surfaces – shapes that could not be defined using convenient mathematical equations, and without which structural analysis was impossible. By precisely surveying the forms of the models – including the challenge presented by soap bubbles up to a metre across – the geometry could be described accurately enough to calculate loads and deflections of the full-size structure, and to predict its behaviour with sufficient reliability, thus providing the necessary confidence to build.

Ted quickly came to see the great potential in this approach and was able to offer to Frei Otto an engineer's understanding of what might be possible and how physical models could furnish the analysts with a form-finding tool – the link that was otherwise missing in their chain.

Throughout his time leading a busy practice, Ted continued to pursue his love of exploration – and no one was a more suitable partner than Frei. They knew each other for nearly thirty years and the match was a rare one. Each found in the other an ideal complement to himself; each could help the other to

Opposite, clockwise from top left: Soap bubble foam structure; Soap film between threads, model for the four-point tent; Soap film in a dumb bell plan; Two domes linked by a saddle; the shapes in the plaster model are formed by reinforcing and articulating with ropes and nets and by anchoring at the low points; Suspension models to find the form of cross vaults; Bone structure of the beak of a black stork; Typical three-dimensional drawn nets; Suspended nets as a model for grid shells.

Above, top: Weaver bird nests: Bottom; A spider's web.

OF THE BOARD
OF GOVERNORS

achieve what either alone could not. Both men had a remarkable imagination, but focused on different aspects of the world of materials and structures – one was more geometric, perhaps; the other more statical (but then statics is really only a branch of geometry tempered by the material properties of the real world). They were both fascinated by the shape that structures ought to be, in order best to exploit the laws of physics and the properties of different materials – Frei from the point of view of the academic interested in knowledge and its structure, Ted as someone who wanted to do something with the knowledge. Each was a great student of that definitive study of structure, form and materials, *On Growth and Form,* by zoologist D'Arcy Wentworth Thompson, first published in 1917, a seminal work (though much neglected) in its deep appreciation of the structure and forms of nature. Frei and Ted felt strongly that only by model making was it possible both to gain a direct appreciation of the scientific equations describing nature, and to develop an understanding of materials, structure and form that made it possible to create *new* structures, not copies of those which already exist. Their friendship knew no boundaries no budget constraints ever frustrated the total involvement of both men. It was as if they instinctively knew that each problem offered intellectual stimulation that was in essence personal sublimation.

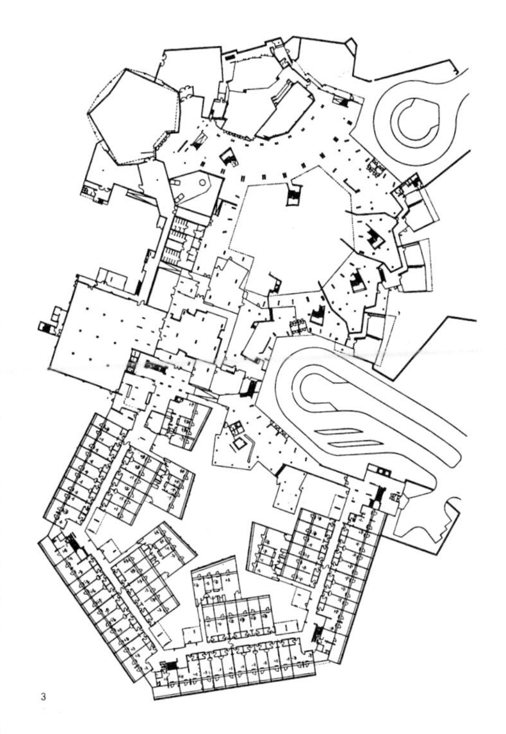

What united Frei and Ted was their wish to explore the unknown and to push back the limits a little. In fact this approach is all too rare among engineers, especially in the building industry, where more often than not they will say that something is impossible if it does not fit the usual range of solutions. There are very few engineers indeed, such as Ted Happold, who have managed to walk the tightrope – to rise to a challenge, to be ready to undertake research without the guarantee of the necessary results, to be able to use their understanding of materials and structures to conceive structures jointly with architects.

If Frei Otto was the inventor manqué to Happold's facilitator, Rolf Gutbrod was equally important in a very different way. He brought to the table a sophisticated expressionist fervour. He was perhaps the German equivalent of the Wolfitt–Olivier tradition of actor manager. He could charm the birds off the trees, advise Arabian monarchs on hip-replacement technology and conjure up exotic and unusual entertainment at the drop of a hat. His method of working – from intuition and flair rather than logic and evaluation – was also perhaps a little wilful in the eyes of an engineer. He could generate a ground plan from a paving pattern or an enclosure from the notation of a concerto. He exuded style, and towards the young Happold in 1967 he also demonstrated shrewd Swabian judgement. He trusted him to deliver Mecca as conceived, and from that date was the staunchest of patrons. Gütbrod over-ruled Ove Arup's objection to include the embryo Buro Happold in the team for Kocommas. There is no doubt the working relationship forged on Mecca with Rolf and his friend Frei Otto was pivotal to Happold's success. The sybaritic side of Ted's nature admired the Steiner-educated Gutbrod's panache and resolve and Otto's messianic obsessions and ceaseless experimentations. Their ambitions were mutually compatible and Mecca was important in that Happold's first exercise in the lightweight tradition was a resounding success – aesthetically and culturally.

The formal world of gravity-suspended roofs, which are also known as heavy tents, is very varied. The pagodas and temple roofs of the Far East are made of fully suspended nets, originally of bamboo lattices. The form studies with models made of suspended chains by Frei Otto and his team revealed the for-

Opposite, clockwise from top left: The Conference Centre at Mecca (the wooden sun protection Kafesses lie on a large mesh lattice of steel cables); Detail of Kafesses; View of the whole complex; Interior of the large Conference Room.

Above: Drawing of Mecca Conference Centre.

BAGS
FOR BASILDON
SUNNY
MARTINS
34

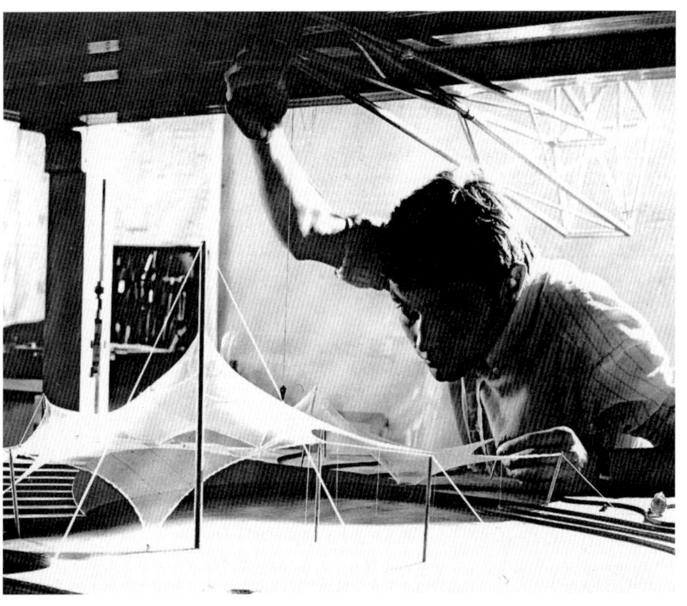

mal world of historical suspended constructions, but they are also key items in the design of suspended constructions made of nets, ropes and tension-loaded bars.

In fact, the final roof at Mecca was not a pre-stressed, double-curvature net as had been proposed by Gutbrod and Otto in their competition entry, which drew on their experience at Montreal and Munich. Rather, it was a suspended cable roof with single curvature, like the two Saarinen buildings at Dulles and Yale. With a span of some 46m it is to this day still the lightest heavyweight roof of its type. An intriguing aspect of the project was that, being at a holy site, the building could be visited during construction only by Rolf Gutbrod, who is a Muslim. Otto and the Arup design engineers had to rely on what could be seen using remote video cameras.

Several gravity-suspended roofs span the roofs of the Conference Centre in Mecca. The complex consists of two areas – a hotel and conference centre, and a mosque – arranged in such a way that each surrounds an inner courtyard. The suspended roofs are made up of supporting steel cables and structural angles, and a three-layer sandwich of timber, insulation and a corrugated aluminium roof. The four conference rooms are connected by a shaded perimeter courtyard. Ropes are stretched between the conference rooms and the radial rope trusses carry shady lattice elements onto a tightly curved inner loop cable on intermediate support posts.

The next round of work between Arups and Otto was very much in the medium pioneered so forcefully by the Otto–Stromeyer combination: the sophisticated tent structure. Tents are among man's oldest built construction. They have been used as accommodation by many cultures for millennia. Tents are stretched, plane, load-bearing constructions made of sheeting, woven fabric or nets. The structure consists of one or more compression supports and tension-loaded membranes. Frei Otto's tent construction research after 1950 started a new overall view of this structural form. He was the first to examine the link between form and structure and thus discovered the significance of the self-forming minimal surface for the design and shape of tent structures. He invented and developed models and methods in which correct tent forms are produced by self-formation. The key to achieving a practical tent roof was to use the models to investigate different configurations of boundaries and mast supports, while always ensuring the membrane had sufficient double curvature.

Ted came to be no less passionate than Frei Otto about the importance of making models as part of the design process, and in 1973 Arups and Otto formed the Lightweight Structures Laboratory as a specialist group within Structures 3, based in the firm's Soho Square offices. Ted gathered together a good team of individuals with skills to complement Otto's fine team in Stuttgart. They included Michael Dickson, Ian Liddell, Vera Straka and Peter Rice.

In 1975, under Ted's direction, Michael Dickson and Rod Macdonald of Structures 3, working with Frei Otto, provided a temporary structure for the 1,000 or so people who would be attending the opening ceremony, at Dyce near Aberdeen, of the BP Forties Oil Field. With the total time for design and construction at ten weeks, the design process had to be very streamlined. A humped tent was developed in order to give adequate rigidity in the windy climate, first with Frei using 1:200 and 1:100 models for form finding and

Opposite, clockwise from top left: Covering for Basildon Town Centre (Michael Hopkins, Buro Happold); interior of Humped tent at Dyce, Aberdeen 1975 (DRU, with Frei Otto, Ove Arup & Partners); Design model for Elizabeth II festival tent in Sullom Voe 1981 (Frei Otto, Buro Happold); Modelmaking in the design of lightweight structures.

Above, top: Humped tent at Dyce Aberdeen for opening of BP Forties Oil Field; Bottom: Washington Music Tent (FTL, Buro Happold).

240 m
1000 m
r = 2200 m
0
500
2

geometric patterning, and then using quite simple calculations based on the experimental studies to increase confidence about its structural behaviour. Both constructions were designed for frequent pitching and striking, like circus tents. The Aberdeen tent was later pitched in Hyde Park and covered an area of 3,300 square metres. The membrane was polyester-reinforced cotton on steel and plywood lattice mushroom elements on tubular posts.

In 1976 Theo Crosby of Pentagram had the task of providing a tent for the British Genius Exhibition in Battersea Park and was able to provide one of the very first jobs for Ted's new practice. Ian Liddell, who had been among Ted's staff at Arups and been involved in the Dyce tent, took the idea of the humped tent further, combining it with a mast and cables to create a large column-free area beneath. This experience became more and more sophisticated, culminating in the extraordinary project (with Michael Hopkins) for the covering of Basildon's town centre – a project fully designed and detailed, but, alas, never built.

A further category of structures that find their own shape are those inflated by air. An English motor car manufacturer, Frederick William Lanchester, first recorded the idea of supporting tents by internal air pressure in 1910, but the real development of air halls occurred in the fifties with Walter Bird's work in the USA. Bird was initially an aircraft and aviation engineer who was interested in the problem of constructing housings for America's exclusive radar protection programme in the north of the continent. Bird set up his own firm, Birdair, which soon became a world leader in the manufacture of air halls and tents. Frei Otto was also working on inflating aluminium membranes as early as 1952. He built tents for Stromeyer and Zelta and observed that pole supported tents can stand without poles if there is increased interior pressure. Many early air halls were created without a great deal of basic knowledge – for example, without considering the extensive leakages that occurred during severe storms and heavy snowfalls. However, halls were improved step by step and large-capacity air halls were built in the most difficult climates. Ideas for building large rope nets under which residential and working space could be created in inhospitable areas, which Otto had already sketched out in *Das hängende Dach*, were brought close to realisation using the air hall in building applications was divided with air at both low pressure and high pressure.

In 1970/71 Frei Otto's Warmbronn studio, Kenzo Tange and Ted Happold's group in Arups (Structures 3) were planning a low-pressure air-inflated roof to cover an entire new city in Antarctica. It was to span 2km and provide a controlled climate for 40,000 people. Engineering knowledge of how air supported structures performed was inadequate – as Ted Happold discovered when he sat on Frank Newby's panel examining British Standards in Air Supported Structures in 1974. In 1978 at the University of Bath he organised a research group through the Wolfson Foundation which related the technology of air supported structures to building performance and process.

One of the successes of the Wolfson Group was that it was driven by engineers experienced in the needs of real construction. They in turn were supported by top-level researchers. The former included Ted Happold, Ian Liddell, Michael Dickson, David Wakefield and Dr Michael Barnes the experienced analyst of the City University. The research group comprised Professor Bryan Harris, a distinguished materials scientist; John Howell, an aerodynamicist, and Chris Williams, a mathematically gifted engineer, both of whom had worked with

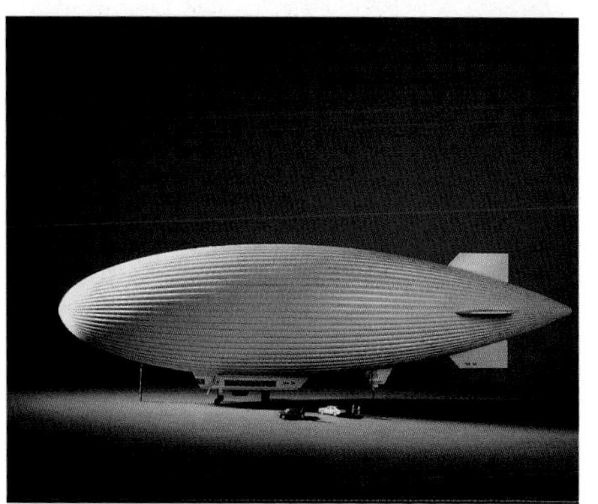

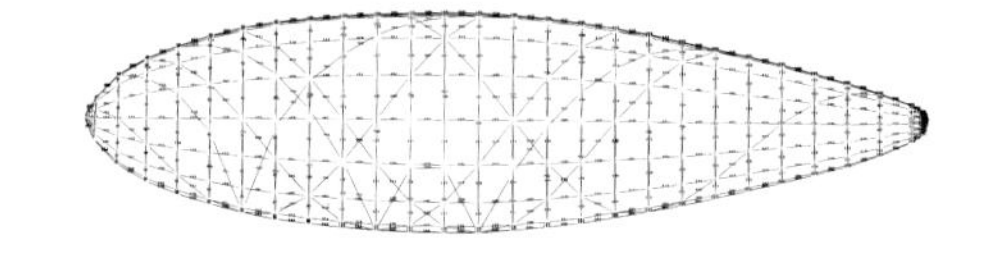

Opposite top four images: Town in Antarctica (Kenzo Tange, Ove Arup & Partners, wirh Frei Otto); 1971 project study 1971 for an air hall as a protection against climate over a residential town spanning 2km, a project a forerunner for 58∞N carried out later with Arni Fullerton and Buro Happold; Below: Models in plaster and perspex; Bottom: Model study for Airfish (Supporting bodies for non-rigid airships 1978, 1979, 1988).

Above, top: ETFE foil cushions used for roof cladding on Chelsea and Westminster hospital (Shepherd Robson, Buro Happold); Middle: High-speed airship to travel at 185kph. In order to handle the vastly increased internal pressure a new construction system had to be originated using an internal network of high-performance Kevlar belts; Bottom: Computer model for airship analysis (Buro Happold).

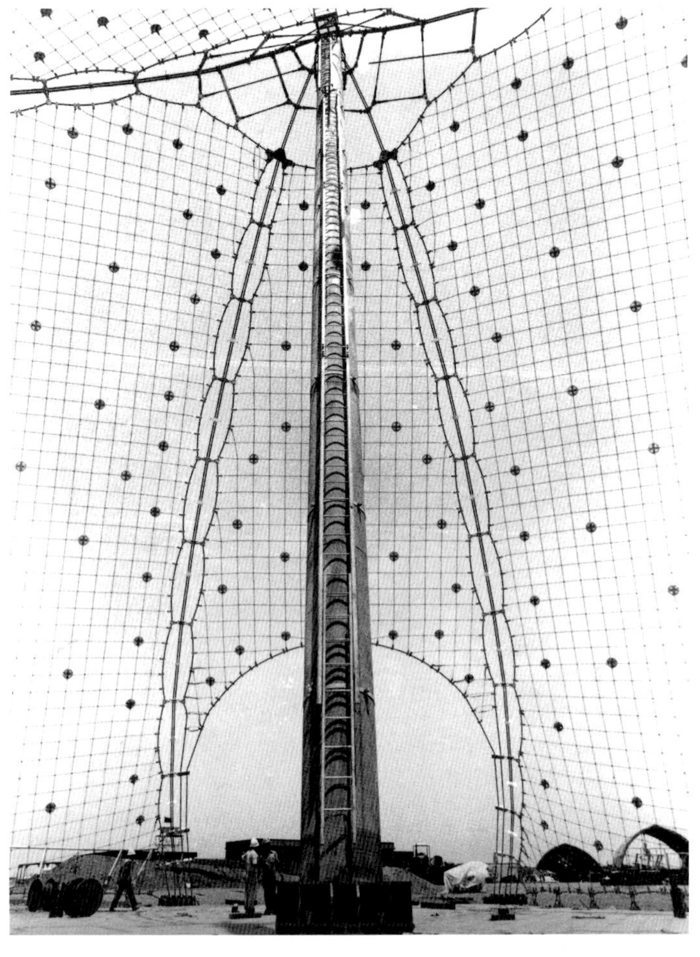

Ted on Mannheim; Dr Derek Croome, an environmentalist; and Dennis Hector, an American architect whom Ted had known at Frei Otto's Institute. This group, under Ted's guidance and direction, was responsible for the engineering totality of two substantial engineering conferences: 'Air Supported Structures – State of the Art' (1980) and 'The Design of Air Supported Structures' (1984), both published by the Institution of Structural Engineers.

In parallel with this research work Buro Happold set about developing software for form-finding and the analysis of tension and fabric structures. This Ted perceived as being the key to the development of these structures and success with the software 'Tensyl' led to Buro Happold and FTL Happold becoming one of the world leaders in this technology.

In 1981 Buro Happold was asked to advise Arnie Fullerton on a smaller version of Arctic City called 58ºN. Ted Happold, Ian Liddell and Frei Otto were able to devise a means for using cables to stabilise the air-inflated dome and to develop a skin for it which was itself an inflated structure – a series of ETFE foil cushions that would provide an excellent heat barrier. This idea for cladding has been used in several recent Buro Happold buildings (Chelsea and Westminster Hospital, Schlumberger in Cambridge, and the Eastleigh tennis courts).

Ted Happold's continuous dialogue with Frei Otto covered another mutual obsession – the shape and structure of pneumatically supported airships. This culminated in the 'Airfish' project study, with Ted Happold, Ian Liddell and Frei Otto working together examining various models with particularly favourable aerodynamic shapes. The supporting flight bodies of the non-rigid airships Airfish 1–3 consisted of flexible membranes with the parts built of rigid elements much reduced. The tail fins are still rigid for Airfish 2, with the fins are flexible, and for Airfish 3 (designed as a hot air ship) the cabin was intended to be a flexible structure. After years of preliminary work the Institute of Lightweight Structures, with the co-operation and collaboration Bath University and Buro Happold, published the *Lufthallen Handbuch* in 1982.

Cable net constructions are subject to the same laws as membrane tent constructions but can cover considerably larger spaces. The support structure for the King Abdul Aziz University Sports Hall in Jeddah stands over 27m high and covers an area of 9,500 square metres. Though it follows the line of structures dating back to Frei Otto's earlier collaboration with Peter Stromeyer at Lausanne in 1964, the Montreal pavilion in 1967 and the Munich Olympic construction with Behnisch and Partners in 1972, the skill of co-designers Buro Gutbrod, Frei Otto and Buro Happold made the project unmistakably Arabian in concept. The vast and beautiful tension structure, despite its scale, is strongly reminiscent of a traditional Bedouin tent.

The early development of the geometric form of the tent surface was carried out at Frei Otto's Atelier and at the Institute for Lightweight Structures in Stuttgart. The first concept forms of the heavyweight roofs had been defined by rough hanging chain models and when the transition to lightweight prestressed double curved cable nets was made, soap bubble models were studied to gain insight into the form of the minimal surface. It is worth noting here that this design was being carried out at a time when the numerical and computing techniques for the analysis and form-finding of prestressed cable net and tension fabric structures were in their infancy and only labour-inten-

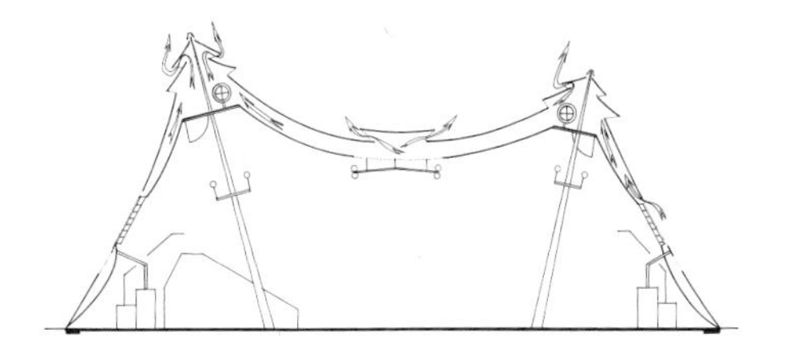

Opposite, clockwise from top left: King Abdul Aziz University Sports Hall, Jeddah, exterior of completed sports hall; Cable net in finished position; Working on clamping elements; Laying out the cable net, masts and cable net in position; the connection of the net boundary cable to the edge, a particularly important detail; Flexibility was achieved by means of electric resistance welded links coupling the boundary cable clamps to the body clamp mounted on the ridge stay cables.

Above, top: Kuwait Sports Centre model (Kenzo Tange, Ove Arup & Partners, Frei Otto); Middle: Soap bubble model for Jeddah; Bottom: Environmental section. To give the tent envelope the improved insulation properties necessary for an air-conditioned space a secondary inner membrane was included which serves both to produce the desired insulation air gap and as a decorative ceiling and wall surface. This membrane is also fabricated in a PVC-coated polyester fabric with 50% translucency.

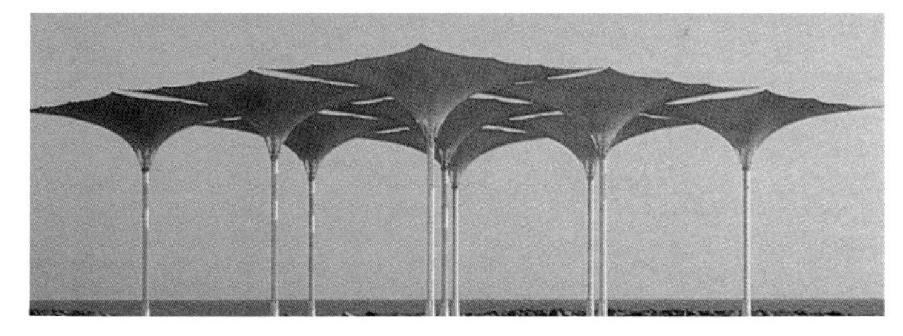

Top and left: The Prophet's Holy Mosque, Medina (Bodo Rasch, Frei Otto Buro Happold), where 17m x 18m umbrellas in the inner courtyard provide cooling shade for the pilgrims; Above: Prototypes on the Red Sea Coast.

Opposite, from top: Pink Floyd umbrellas 1978; Shade in the Desert for the Hadj (Rolf Gutbrod, Frei Otto and Ove Arup & Partners) 1972; Convertible roof Cabrio folding stand cover 1986 (Frei Otto, Buro Happold).

sive main-frame analysis systems were available.

Production of the steelwork components took place during the latter half of 1978 and they were shipped to the site for a start on the cable net erection in January 1979. The masts were transported to site in segments and the first site operations involved welding up the 30m-long masts and the installation and casting-in of the mast base sand pots. Following weld certification testing, temporary erection steelwork was assembled at the mast heads, which were then craned into position. These were placed (at their approximate final inclinations) into their respective sand pots and guyed off with a system of temporary guys determined by earlier method studies. The assembly of the cable net fields and boundary cables then proceeded. A comprehensive cable numbering and marking system had been decided upon during production of the cable schedules, and this proved invaluable on site, both in allowing cables to be sorted into their correct location and orientation with minimal abortive effort, and in enabling errors of manufacturing to be identified at the absolute earliest opportunity. The complete laying out and clamping took approximately eight weeks for the 10,000 square metres of net, which included approximately 40,000 cross clamps, 3,500 boundary clamps and 500 node and stay cable clamps. The entire net assembly in the air weighs approximately 8kg per square metre, with a further 8kg per square metre of ground plan of steelwork in the masts and anchorages. The primary lift took approximately seven days in May 1979. During this time the tangled mass of cables bundled on the ground unfolded into a graceful, light, metallic, orthogonal spider's web. During the design development on the Jeddah project and the Munich Aviary, new methods of mathematical and geodetic calculation to complement model building were developed at the Institute of Lightweight structures.

Models at a scale of 1:125 were made according to Happold's preliminary calculations of forces in the net supports and guys and using geometrical data from the design models, the detail design of the cables, masts and boundary supports could be completed. The increasing sophistication led to the development of various aviary forms by Jorg Gribl, Frei Otto and Buro Happold. These nets had a very fine mesh grid of 62mm x 62mm and, in the case of Munich, span a high interior, covering 4,600 square metres supported by ten masts with nets suspended from self equilibrating clamping plates. These nets are woven in crimped stainless steel wire and are able to withstand a maximum snow load of 38kg/metre square.

One aspect of the art of design which particularly suited Frei Otto and Ted's experimental approach was that of turning an idea that is both obvious and easy at a small scale into something that is many times larger and needs to be much more durable. It was a great tragedy the Otto and Happold concept for Shade in the Desert, the temporary structures for the Hadj, was not realised or that the Cabrio convertible construction did not go beyond the prototype in 1981.

In 1977 Ted Happold became involved in two further stages of developing Frei Otto's remarkable concept for umbrellas and parasols, which had begun in the fifties and reached a high point in 1971 with the 19m umbrella for the Garden Festival in Cologne.

Ted and Frei worked on a group of similar umbrellas for Pink Floyd's 1977 tour of the United States. The umbrellas, made of white cotton and 4.5m in diameter, had to rise out of the stage and unfurl as pristine lilies. Then – in the early ninties, with Bodo Rasch, who had worked with Frei Otto on the Cologne and Pink Floyd umbrellas – the idea was taken a stage further and folding parasols were used to control the entire environment within the courtyard of the Prophet's Holy Mosque at Medina. Rasch invited Buro Happold, represented by Eddie Pugh, to join the team which developed a suitable construction for the parasols. Because the fabric needed to be capable of folding and unfolding thousands of times without damage, a fine Teflon (PFTE) fibre was used to create a silk-like material, Tenara, with extreme pliability and total resistance to fire and to the intense ultraviolet radiation it would be subjected to. The fabric would also need to be protected when the parasol was furled, like the shell which protects a folded insect's wings, and for this a series of flaps made from carbon fibre laminate was developed. Integrated within the cast-steel support for the parasol itself was a sophisticated system of hydraulic actuators, supply ducts for cold air circulation, loud speakers, and lighting – a triumph of interdisciplinary collaboration.

The King's Office and Council of Ministers, Majlis al Shura (Kocommas) was one of the most ambitious projects for the Kingdom of Saudi Arabia. It was to be the design for a Government campus on one square kilometre adjacent to Riyadh's established Diplomatic Quarter. The project started in September 1974 when letters of requirements and function for the King's private office and for the Council of Ministers were given by the Minister of Finance to Rolf Gutbrod. When initial designs were presented to the Saudi Arabian Government by Rolf Gutbrod, Herman Kendel and Ted Happold in November 1974 the design team was asked to extend the scale of the project by adding the Majlis al Shura, an assembly for a Government advisory body. The design team was initially a joint venture between Buro Gutbrod and Ove

Opposite, top: Site model for the King's Office and Council of Ministers Complex, Riyadh (Rolf Gutbrod, Ove Arup & Partners, Buro Happold); Bottom: The early model of the Majlis al Shura which led to some highly sophisticated development studies of one of Frei Otto's obsessions – the use of a branched structure as support for a hexagonal lattice shell.

Top: Shades or feathers for the octagonal shells for the Majlis; Middle: A view through a development model; Bottom: further development of the shading devices for the Majlis al Shura.

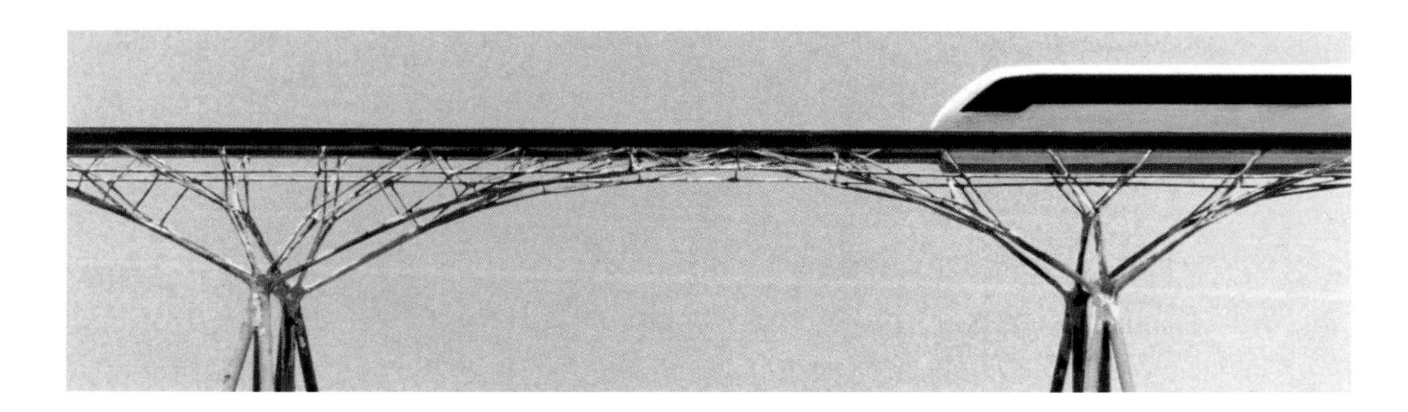

Above: Magnetbahn, alternative structures to carry the train; Opposite: Michael Dickson and Ted Happold on the test site, and other development models.

Arup and Partners supported by many specialist consultants, including amongst others Frei Otto (lightweight structural forms), Schmidt Reuter (building services), Buro Lutz (landscaping), and Widnell and Trollop (quantity surveying). When Ted Happold and his partners established Buro Happold in May 1976 Rolf Gutbrod insisted that Buro Happold and Ove Arup and Partners should jointly carry out the civil and structural engineering design work and the joint venture was enlarged to include Buro Happold. After the death of King Khaled, despite the completion of a substantial initial works contract, the project was aborted. This was a tremendous blow to the participants but two factors of major importance came out of it. It acted as a catalyst and key factor in the early growth of Buro Happold and provided support for valuable engineering research into the highly complex roof structure of the Majlis al Shura, a variety of shade structures and the translucent marble façade of the Council of Ministers building.

Magnetbahn

In 1991 the Federal Ministry of Research and Technology commissioned Frei Otto and Ted Happold to prepare a report on the feasibility of alternative designs for the German high-speed Magnetic Levitation Railway system. The brief was to research various means of viaduct construction to that used on the existing test track in northern Germany, which had been built using massive post-tensioned concrete box beams on substantial reinforced concrete frame columns. The development of the aluminium coaches for the Magnet Express train was already advanced and had been tested to speeds of 450km per hour. However, the structural form and resulting aesthetic and ecological impact of the raised track not only played an important part in the feasibility of the whole magnetic levitation train project but also comprised over three quarters of the

total cost of the system. To accommodate the high speeds attained by the train, considerable stiffness of the viaduct bridging system was required in order to maintain the tight tolerance of the magnetic gap. Despite the fact that the train itself weighed relatively little, it was especially important to search for a bridging system that would distort only minimally under the effects of high temperature in the sun and under the influence of the train itself. The new track had not to blight its surroundings and had to fit in visually with the landscape in such a way that it could form an inconspicuous part of the ecological system. It had, therefore, to be extremely light and exert as little force as possible on minimal foundations. It had to be easy to assemble and not disturb the environment during erection or thereafter by forming a barrier preventing cross-migration beneath from roads, water courses and by animals.

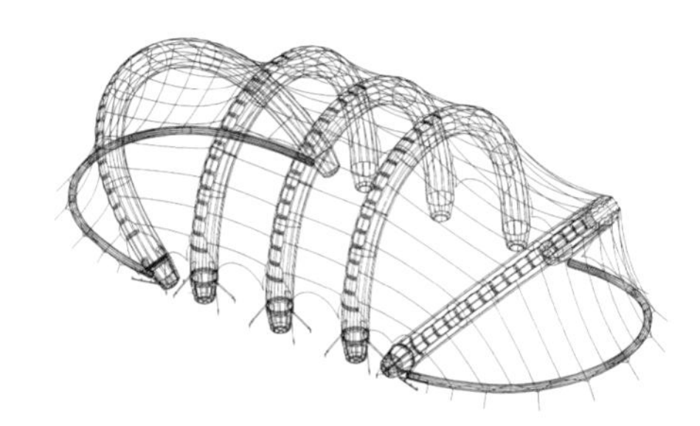

The study examined ten different structural forms, and calculated and optimised a number of structural variants. The study included physical form-finding methods for determining direct force path systems and various forms of branching structure. All designs were assessed to meet the stringent aesthetic, ecological and structural criteria set by the Ministry of Research and Technology. The models exhibited optimised the sixty variants considered and show options for the working path from the solid concrete or steel beams to the filigree lattice beam in corrosion proof steel intended to be constructed as a static-dynamic continuum.

As well as wanting to pass on responsibility for design to younger generations within the practice, Ted Happold had for many years been keen to have a more established presence in America. Since the early eighties Buro Happold have been carrying out the fabric engineering for the New York practice Future Tents Limited (later FTL Associates), among whose founding partners were Nick Goldsmith, Dennis Hector (who had worked with Frei Otto in Stuttgart) and Todd Dalland. Hector also helped Happold start the Lightweight Structures Laboratory at the University of Bath.

One of the first major projects was the Baltimore Harbor Lights temporary canopy in 1981, since replaced by the Pier 6 Concert Pavilion. Another early FTL project for which Ian Liddell gave advice was the Carlos Moseley Music Pavilion. The commission was to provide a mobile stage for free public concerts given by the New York Philharmonic and the Metropolitan Opera in sixteen separate park locations throughout the city, to be erected or dismantled in about six hours. The polyester tensile canopy is supported from the apex of a pyramid formed by three hydraulically-operated unfolding steel struts. The canopy itself is curved, not only to provide adequate protection against rain for the performers but also to project the sound more effectively to the audience.

After some ten years of collaboration with FTL Ted Happold realised that there might be mutual benefit in creating a joint practice embracing both the architectural and fabric engineering aspects of tension structures. This culminated in the formation of FTL Happold in 1992. It comprises roughly equal numbers of architects and engineers, and among the many other projects the group has recently undertaken is the deployable tension structure for the Cadillac Theatre and, more recently and using the group's expertise in low-pressure air-inflated domes, the provision of a fully transportable maintenance building for the US armed forces' helicopter fleet. It has also developed a high-pressure version in which tubular arches are inflated to support a loose fabric structure providing covered maintenance areas.

Opposite: Work with FTL – The Baltimore Harbor Lights, Pier 6 Concert Pavilion. The three bays covered by the membrane structure provide space for 3,400 seats. The outer promenade accommodates an additional 1,000 patrons.

Above, from top: A high-pressure air-inflated fully transportable maintenance building for the US armed forces helicopter fleet, with tubular arches inflated to support a loose fabric cover; The Boston Harbor Pavilion; The Hyperion Airship developed for the Disney organisation for Disneyland, Paris.

Islam – A Journey

It is important in addressing the body of work in the Islamic world over the last thirty years involving Ted Happold to use his own words to describe the friendships, cultural associations, diversity of projects and personal satisfaction that he gleaned from this extraordinary journey, which will now be continued by his partners who supported his endeavours and shared his ambition. In a sense his combination of gregarious enthusiasm and the Quaker capacity for reflective contemplation was perfectly suited for his introduction to Saudi Arabia, as was his ability to live off the land, assimilated in his early exposure to responsibility as a tour leader for his parents' creation for international friendship and understanding, International Tramping Tours. D.W.

Ted gave the following account at the IABSE Conference on 'Places of Assembly', Birmingham 1994.

It was King Faisal's liberal policy that led to the Saudi conference centre competition. An attempt had been made to hold an international medical conference in Riyadh and the facilities had been totally inadequate. The King decided that increased world participation meant a need for conference facilities. He would link that need with setting a higher standard of building. The United Nations' planning adviser was Dr Omar Azzam, a well-related Egyptian architect, and the Saudi government instructed him to ask the Union Internationale des Architects to organise an international competition and they in turn appointed Theo Crosby as the technical adviser.'

Trevor Dannatt and Ove Arup and Partners submitted an entry and we half forgot about it until a cable arrived, rather to Trevor Dannatt's horror, asking us to build the Riyadh centre. It turned out that we were going to come third, and then the judges asked the King to visit the exhibition. He walked in the room, looked at the models – ours was particularly stunning – put his hand on ours and said he would have three of them. The judges spent the rest of the day trying to change the King's mind! A final compromise was that we would do Riyadh and one other would be built in Mecca, designed by Rolf Gutbrod and Frei Otto.

On New Year's Day 1967 Trevor Dannatt and I went out, via Beirut, and started agreeing a contract – hard work without typewriters and with no means of telephoning or even writing home. We were there three weeks, and in the middle of it, Rolf Gutbrod arrived with two of his staff, Hermann Kies and Hermann Kendel, and they were only too happy to find us working. Obviously many of the problems of the two projects were the same and when Rolf Gutbrod and Frei Otto asked if I would also be their engineer it was agreed, and this really formed two very closely-knit design teams.

The two projects together were very demanding. Fortunately Ian Liddell had just rejoined Ove Arup and Partners and led the Riyadh team, whilst Ting Au ran Mecca. Peter Rice came back from Australia via the USA to work on the roofs for Mecca. Both designs were unusual, Mecca so much so that it subsequently won an Aga Khan award as the most technically innovative building for a decade in the Islamic world. Apart from the standard of design though, the interesting aspect was the relationship with the clients, for whom it was an entirely different scale of project and one which represented an ambitious start to Saudi Arabia's modern era of construction.

Opposite, clockwise from top left: Two interiors of the Intercontinental Hotel showing the high quality of the concrete structure (Trevor Dannatt, Ove Arup & Partners); The main entrance to the Conference Centre in Mecca by Rolf Gutbrod with Frei Otto.

Above, from top: A typical view of the Hadj, the annual pilgrimage to Mecca; Competition entry for Shade in the Desert which preoccupied Frei Otto, Rolf Gutbrod and Arups for many years.

Both projects went out to tender in 1968 and COGECO – an Italian firm headed by Otto Vannucci, whom I had met when studying the Italian building industry for the DOE when working on the British Embassy in Rome – won the Riyadh project. Together with the architects we had the responsibility of supervising the work on site. Only Muslim engineers were allowed into Mecca and there were not so many available at that time. There had been talk of supervising by television but common sense prevailed and a site office for both the contractor and ourselves was set up at Hadda, on the border of Mecca. A partnership had been formed between a Lebanese firm of contractors, Joseph Khoury, and the French firm of Thinet, but the latter had to take over most of the work and employed some very competent French Muslim workers.

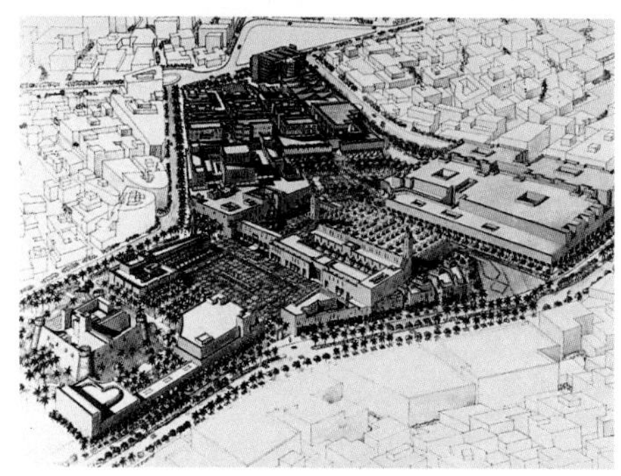

The projects had been envisaged as conference centres with government guest houses attached. However, practicalities of the operation meant major hotels became involved and they were an instant success and were immediately enlarged. The conference centres became a lesser adjunct.

The Saudis are very conscious of living in a harsh desert environment and this is an important influence on relationships between people. When they like you they want to discuss things with you all the time. As the government advisors and officials started to understand what we could achieve, they wanted introductions to other western architects. Trevor Dannatt introduced Sir Leslie Martin, who designed a major government centre for the mountain town of Taif for use in the summer, and Sir Leslie used our ideas for massive 'mushroom' shade structures where the services ran within the columns and the soffits reflected thermal movement.

It was a wonderful time enhanced further when Professor Karl Schwanzer, an Austrian architect and planner, asked if we would be interested in being considered for all the engineering, costing and programming aspects of the master plan for the University of Riyadh, which had been given a two mile by one mile site on the edge of the city. We did the management and civil and structural work, and sub-consulted the building services and surveying.

We did not continue with the work after Karl Schwanzer died, but during that period we met a whole generation of gifted and well-educated young Saudi engineers. Some of them have worked with us, and some we met later as our clients. The knowledge and experience was building up and there is no doubt that the very broad basis of consultation often achieved a very high standard of work.

One of the more interesting projects just after this was a limited competition for the replanning of the area around Mona for the Hadj, and annual pilgrimage. Oddly enough Rolf Gutbrod and Frei Otto had been asked separately but there was great integration between the two. Here we met for the first time a young Saudi Arabian architect, Sami Angawi, who came with Bodo Rasch from the University of Austin in Texas to work on the entries. Sami comes from one of the families who hire tents for the Hadj, and his experience of the process of the pilgrimage – for most of the way a crowded walk in hot sun, with stops for stoning the devil, for sacrifice, sleeping accommodation and the like – did lead to a fascinating solution. Alas, as the competition was going on another ministry was carrying out major changes and our efforts were abortive. But Sami, helped by Bodo, started the Hadj Research Centre at the University of Jeddah and Buro Happold, just established, were appointed with Arups as engineers for the King Abdul Aziz University Sports Hall,

Opposite, top: Study model for Kocommas, the Majlis al Shura (Rolf Gutbrod, Frei Otto, Buro Happold); Bottom: Kasr al Hokm district of Riyadh, the heart of the original walled city which was losing its identity. An international competition was held in 1984. An international competition was held for the second phase of the City Centre development. This involved three major public buildings – the Justice Palace, the administration seat of the Governor of Riyadh, and the Grand Mosque. Buro Happold were asked to act as engineers and project managers for the public buildings. Shubeilat Badran Associates were appointed architects for the Justice Palace and the Grand Mosque, and Scott Brownrigg Turner for the Culture Centre.

Above, from top: Kocommas aluminium model for the Majlis; an aerial perspective of the Kasr al Hokm.

Jeddah. Arups did the ground works and we did the superstructure, and the cable net solution was then the largest fully enclosed structure of its kind and remains to this day one of the most elegant cable net structures ever achieved. Frei Otto and Rolf Gutbrod were the architects, continuing their partnership established in Saudi Arabia for the Mecca Hotel and Conference Centre.

At about the same time, Rolf Gutbrod was in discussion with the Ministry of Finance for King Faisal's plan for a government complex which would house the King's office, Council of Ministers and Majlis Al Shura (consultative council – in essence, a parliament). I went with him to finalise the design contract, and as Buro Happold was just then being established, we all agreed that the design work would be carried out jointly by Rolf Gutbrod, Ove Arup and Buro Happold, with a number of other international consultants acting as sub-consultants. We referred to the project as 'Kocommas'. This was a bold project, with translucent marble façades and exciting hexagonal grid funicular lattice shell roof structures which we helped Frei Otto to develop. The design was completed, but impetus for the project waned after the assassination of King Faisal, and though a substantial initial site work contract was completed, the project was eventually shelved.

At the time that the Kocommas initial works contract was on site, Trevor Dannatt was appointed by the British government for the design of the British Embassy in Riyadh, and (Buro Happold) were appointed as engineers. The embassy is located in the Diplomatic Quarter of Riyadh, and we had also entered a competition for the Diplomatic Club, together with Frei Otto and the San Francisco firm of architects, Sprankle Lynd & Sprague. We won the competition, but failed to finalise a contract just as British television screened the controversial *Death of a Princess*. We thought that was the end of our chances of being awarded the project, but the client, the Bureau for the Diplomatic Quarter, waited a year, till relations between Saudi Arabia and Britain returned to normal, and they again invited us to discuss a contract with them but now with Omrania as architectural partners. The completed building, now called the Tuwaig Palace and used for government and diplomatic functions, is seen as one of the major landmarks of the region.

Based on the considerable success of the Diplomatic Quarter, the Bureau for the Diplomatic Quarter was appointed planners for the city as a whole, and its name changed to the Arriyadh Development Authority. It set about attacking the whole range of problems which had arisen from the immense and rapid growth of the city. Prime among these was the accelerating loss of identify and importance of the Kasr Al Hokm district, the heart of the original old walled city. The Authority held an international design competition for a number of the most important elements of the city centre.

All the partners of Buro Happold and very many of the staff have at some time or another worked on Saudi Arabian projects, either working from our head office in Bath or in Saudi Arabia. Two partners who have committed large periods of time in Saudi Arabia are Eddie Pugh and Padraic Kelly. Eddie Pugh was resident engineer for the King Abdul Aziz Sports Centre, he was site project manager for the Diplomatic Club, and he later went back to manage the construction of the GCC Conference Centre with the Arriyadh Development Authority.

I have worked in Saudi Arabia for nearly thirty years, most of which have been with the same people. It has been exciting and fulfilling for both me and my colleagues. We have met and worked with so many talented and able people from that country and we have found amongst them many friends.'

Opposite, top: Erecting the mast and cable nets at the King Abdul Aziz Sports Centre, Jeddah. Note the scale of the masts and nets; Bottom: A view of the Diplomatic Club, Riyadh, with the Heart Tent in the foreground.

Above, from top: Trevor Dannatt, Ted Happold and Theo Crosby in Riyadh in the late sixties; The Diplomatic Club in context.

The Grand Mosque, Riyadh

A competition was held for the design of the new Grand Mosque in Riyadh and in 1985 the Arriyadh Development Authority appointed Shubeilat Badran Associates as architects and Buro Happold as engineers for the project. The concept design for the new Grand Mosque was very deliberately in character with the traditional building forms of the Central Province. However, the materials used in the early traditional Nejd construction were not appropriate for a modern building of the magnitude of the Grand Mosque, which lay in the Kasr Al Hokm district, at the heart of the old walled city. The prime objective was to re-establish the area as the cultural centre of the city, visibly reflecting its historic past. In developing an engineering solution for the building, two sets of criteria had to be met. Firstly, the building should clearly follow the forms and principles of construction of traditional buildings in the region. Secondly, air conditioning, lighting and other building services should be incorporated into a structural system in such a way as to achieve a co-ordinated and logical, rather than forced, engineering solution. A precast form of construction was developed for the structure of the Grand Mosque, with columns supporting column heads and a system of double walls above, which in turn supported a precast double tee beam roof. The double walls have the style, form and mass of the early tied masonry arch walls, and the double tee beams give a modern textural equivalent of the early tamarisk pole and adobe roof. The precast double wall not only serves the same structural purpose as the early tied arch masonry, but can span further and provide space within it for mechanical and electrical servicing, in particular for cooled air which flows within the walls and into the prayer hall. Terry Ealey was Ted's most constant collaborator in the Islamic world and his drawing of the Grand Mosque structure is illustrated on the left.

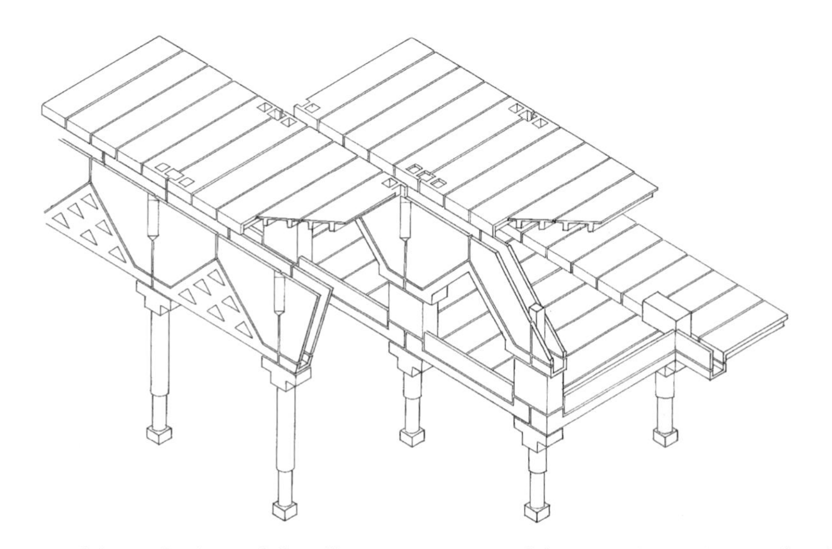

Top: Interior of the completed prayer hall; Middle: precast construction of the exterior; Bottom: Isometric through mosque showing main prayer hall, arch, beam column and roof structures.

Quba Mosque, Medina

The new Quba Mosque, finished in 1986 and lying on the site of an existing mosque built some 150 years ago and restored in 1967, is designed to take up to 10,000 people in prayer. With Mecca to the south, precisely determined by the orientation of the existing Qibla wall, the mosque is located in the Al Menaura district of Medina, on the sacred and historic spot where the prophet Mohammed built the first mosque in the Islamic faith on his emigration to Medina. As an existing burial ground lay to the east and a school to the south, the mosque could extend only to the north. The plan of the mosque, established after a survey of others in the Medina region, follows the traditional form of a rectangular building with a central courtyard. This courtyard provides access to the prayer hall and is shaded by a lightweight retractable membrane, designed by Dr Bodo Rasch for use during prayer and during periods of extreme heat. The north end of the mosque is framed with a pair of minarets, the larger lying adjacent to the main entrance. The entrance porches do not lead directly into the mosque but maintain traditional movement patterns with circulation being offset from the main area. The entire structure above the reinforced concrete and hollow-pot podium is of brick masonry. The 12m and 6m timber domes were structured in load-bearing brickwork without supporting framework and with only minimal reinforcement to control hoop cracking during construction. The four tall, thin minarets at either corner of the mosque are also in masonry construction but post-tensioned through their central core. The quality of traditional detailing and Islamic finishes owe much to the scholarship of the architect Abdul-Wahib El Wakil whose exceptional knowledge of both traditional masonry forms and Islamic decoration is central to the exceptional quality of the mosque.

مجموعة من الصور تمثل مناظر متنوعة للمسجد

Top: Composite details of the Quba Mosque; Bottom: Plan of house, garden, shops and ablutions area.

The University of Qatar

The University of Qatar had a long gestation period. Ted Happold in his role as Executive Partner of Arups Structures 3 was commissioned with an Egyptian architect, Dr Kamal Khafrawi, in collaboration with Renton Howard Wood to develop the master plan of the University for the State of Qatar. The campus was to grow incrementally on a modular basis for a student population of some 4,700. The planning concept is based on an octagonal grid, a traditional Islamic form. It uses high-quality white, precast-concrete cladding and incorporates a traditional Arab wind tower over classrooms to allow indirect natural light and early ventilation through opening lights within the roof structure. The structural design solution is a specific response to the architect's planning grid. The concept for high-quality concrete buildings in a modular low-rise form has allowed the use of repetitive precast elements for both cladding and structural walls. The adoption of precasting had the advantages of rigorous quality control in production and increased speed of construction with a reduced labour force. The planning of the academic buildings was based on two grid forms, an octagon 8.4m in width and a square with sides of 3.5m. The octagons are adjacent and connected to the squares to form the modular grid pattern. For energy conservation sandwich construction has been adopted in external walls made up of facing panels, insulation and loadbearing or infil block inner panels. Primary structure consists of loadbearing precast concrete walls with floors formed of waffle or trough-type cast-in-situ concrete. Roof structure variations occur where two octagons are combined to produce laboratories. Vierendeel precast trusses span the double octagon. Where four octagons are combined, a cast-in-situ folded plate roof geometry has been adopted.

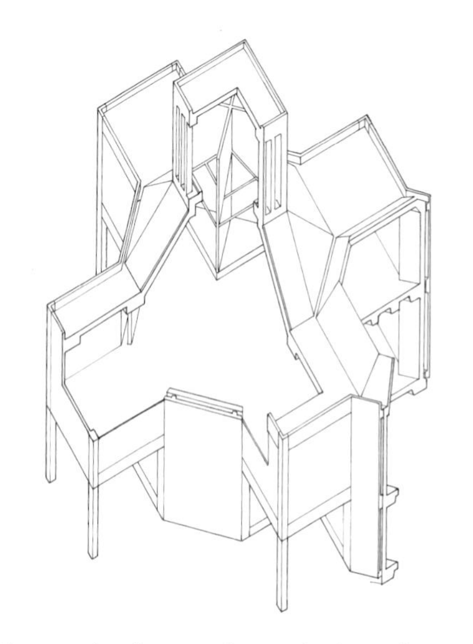

Isometric of wind towers with external and internal views

The Al Marzook Centre for Islamic Medicine, Kuwait

The Al Marzook Center for Islamic Medicine was opened by Their Highnesses the Amir and Crown Prince of Kuwait in 1987. The centre comprises a mosque for approximately 1,800 worshippers, an out-patients clinic, laboratory facilities and library, together with the management offices and a conference auditorium for the Islamic Organization for Medical Sciences. The latter is an international body engaged in the furthering of Islamic (homoeopathic) medicine, was formed in Kuwait by Amiri Decree in 1984. However, the first steps towards establishing this organisation were taken some years before when, in 1978, Khalid Al Marzook and his brothers pledged their support and offered to build, as a gift to Kuwait, not only a base for the organisation but also the other facilities. The IE Zekaria Partnership was appointed as architects with Buro Happold as structural engineers. The design was completed in 1980 and work began in 1982 on a site close to the existing Sabah Hospital. The scheme was conceived as a series of colonnades forming a covered walkway between the mosque at one end, and the library, clinic and conference building at the other. The form of the colonnade was to be a vaulted arch; this and other related arch forms were adopted as a unifying element between the covered walkway and the buildings. It was strange that the load-bearing aesthetic could best be achieved by the use of masonry, but the validity of such a complex arch and vault masonry construction nowadays is doubtful, even in a country where building stone is plentiful. Since there was no natural building stone in Kuwait and sufficient skilled labour was not available, concrete, either precast or in situ, was considered to be the most appropriate material for the development.

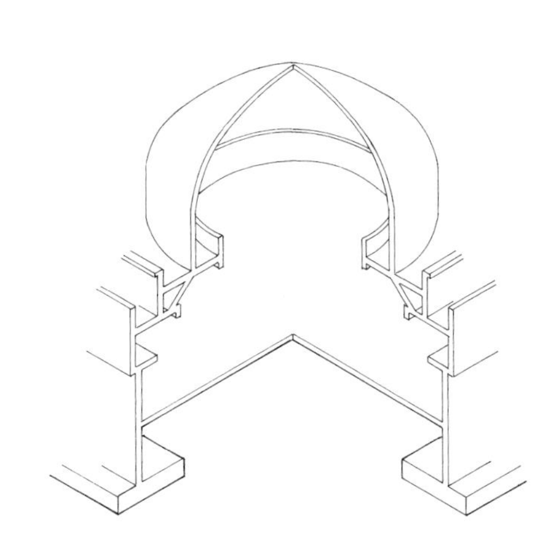

Section through Mosque dome, internal and external views

Above: A model of the Mosque and Cultural Centre and varying stages of construction of the timber framed domes and finished columns.

The State Mosque of Sarawak

Sarawak is a state of modern Malaysia, situated on the northern side of the island of Borneo. For 100 years, from 1840 to 1940, the country was governed by three generations of Brookes – the 'White Rajahs' who in an idiosyncratic, liberal style attempted to control and organise the various indigenous tribes, both dissuading the Dayaks from head hunting and controlling the Chinese traders. The Malays, however, were Muslims and by virtue of their Bumiputra policies have attained political ascendancy. After some consideration of designs by local architects, Sami Mousawi was commissioned to prepare a design for a mosque and cultural centre. Buro Happold was appointed as structural engineers. The concept for the 500m x 500m cultural centre was developed to scheme design stage by Ian Liddell and Ted Happold, with the mosque as the central building. In the event, only the mosque, riwag and minaret went forward to detail design and only the mosque itself was finally constructed with a temporary arrival and ablutions area. The mosque is essentially a square box, 90m x 90m on plan and 10.8m to the general ceiling level. The reinforced concrete walls have a total thickness of 1.8m at ground level, but are heavily modelled and clad externally with reconstituted stone. The roof is supported on complex column clusters on a 10.8m grid, with eight rows of columns in each direction. Each precast concrete column cluster consists of four columns rising from ground level and curving outwards to a collar 9m above. Standing on the collar are eight flying half-arches which support the roof. The smaller domes are hardwood grid shells, while the central dome is a combination of the concrete half arches and infill wood structure enclosed by a timber grillage all to a very high standard of construction.

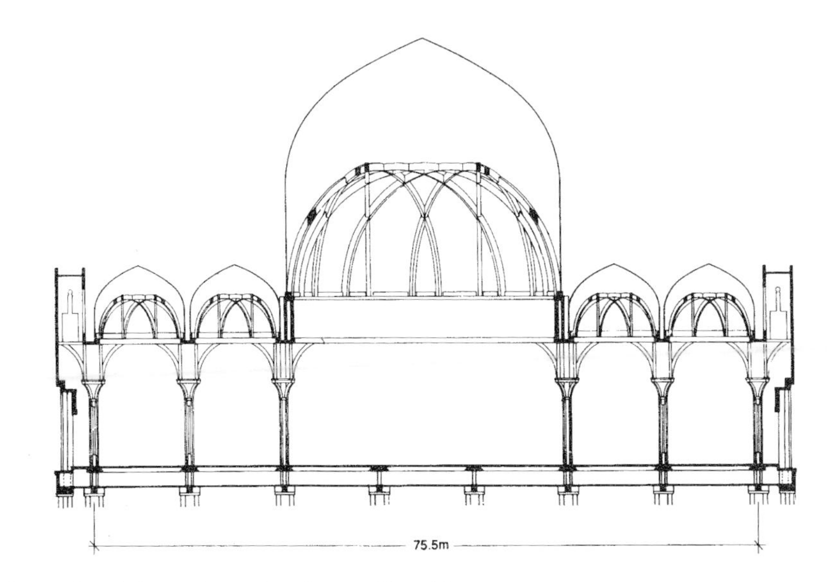

Sections through Mosque dome walls, columns and floor slabs.

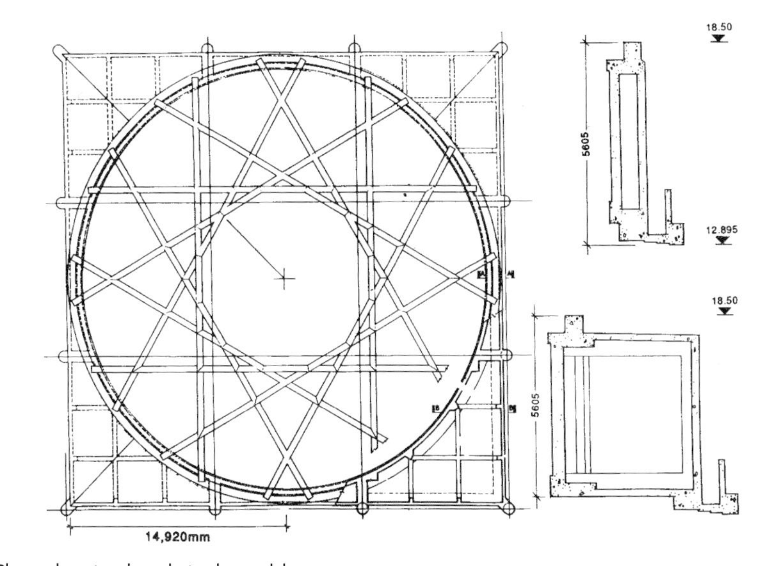

Plan and section through ring beam slab.

View up through column clusters into small dome.

Design Methodology

By inclination, engineers tend to be people who get on with things and get them done. The other side of this coin is that they can be disinclined to reflect on what they are doing when they design structures, and this is why philosopher-engineers are rare. The combination of these two approaches to the world can, however, be particularly potent. Eduardo Torroja was one such engineer. He was admired by engineers throughout the world, not only for his skill in using the full range of construction materials but also for having written *The Philosophy of Structures* (published 1967). This book is a milestone in engineering design history for successfully analysing the many aspects which engineers have to consider when conceiving structures – possible loads, the behaviour of different structural forms and materials, and methods of construction, both separately and together. Virtually alone among engineers who have written about design, Torroja did so by standing back from his own projects and by abstracting general ideas from his particular experience.

Ted Happold was one of only a handful of other engineers who wrote about their art with something of the philosopher's detachment – with Nervi, Ove Arup, Peter Rice and a few others. While their writings make wonderful reading they unfortunately did not, like Torroja, condense their ideas into one book. Nevertheless it is particularly interesting to observe the individual pre-occupations and approaches to design and construction taken by these very different engineers. Contrary to the common perception of engineers as technicians, the results of their work are as distinctive and diverse as the work of different architects.

Ove Arup's particular concern, as a contractor himself, was with the working relationship between architect and both the contractor and design engineer. Way back in the 1930s he foresaw the benefits of such dialogue and collaboration when it was seldom fruitfully exploited. Nervi, too, was a contractor and showed us the engineering creativity that could result from allowing a total understanding of innovative construction methods to inform the conception of a structure. For Peter Rice, it was the outward expression of a material's inherent character that engaged his mind when designing – a material's *engineering aesthetic* as he called it.

Ted wrote on many of these same themes, and more, but unlike some philosophers who have tried to address that elusive, possibly chimeral, moment of conceiving a certain design solution, he was uniquely preoccupied with the actual process of design itself, with the means by which seedling ideas could be nurtured through to maturity. His interest in this was no idle contemplation – from 1967 he was running one of the four design groups in Arups. Not only did he have to ensure that good engineering ideas were fed into design schemes as early as possible (by having highly creative engineers in design teams) but he also had to ensure that innovative ideas could be made to work.

This 'development engineering' phase is rare in the construction industry and few clients are prepared to take the risks associated with innovation. While seldom admitted at the time, clients are frequently not told of the real engineering uncertainties associated with some schemes, lest they lose their nerve and demand a safer, lower-risk solution; and engineers must constantly live

Timber gridshell for Garden Festival, Mannheim. Carlfried Mutschler and Partners, Frei Otto, Ove Arup and Partners. Opposite: Fixing the waterproof membrane onto the timber lattice structure.

Above, top: Frei Otto's suspended chain model to establish the vault forms for the lattice shell. It was not only the design of this structure that was determined by self form processes but the translation of the shape into architecture; Bottom: Typical internal view looking towards garden.

up to the expectation they have generated in past ages: that they can always deliver the goods.

It has to be said that Ted was regarded by many of his peers as over-confident in some of his proposals, even unsafe. In some ways that was true – more than one of his ideas would not have worked in their *original* conception. But this is true of many preliminary designs, especially innovative ones, both in construction and elsewhere. Airlines, for instance, make firm commitments to buy new aircraft and the aeroengines that power them long before it is known *how* the promised performance and cost will be achieved.

But Ted was too shrewd to be foolhardy; he had his safeguards. He knew that he had enough good engineers in his team to ensure no mistakes would creep through; he knew the difference between ideas that could never work and those that would, given the time and skill to develop a design; he had enough confidence, based on experience, to know what would probably work, and what might not; he knew when it might be prudent to have an alternative, safer idea in the cupboard should an adventurous idea prove uneconomic or impractical. This was how he was able to offer to both architects and building clients the confidence to proceed, first to design and then to build.

Much of the early work in Structures 3 was relatively straightforward but soon – inevitably perhaps, given the creative engineers in the group – structural designs were being proposed that were anything but straightforward. Some, such as the large space-truss at the Riyadh Hotel and Conference Centre, were at the limits of what could be calculated using the analytical methods and the relatively small computers available at the time; others, such as the cable roof at Mecca, were exploring new types of structure with no established design procedures and little precedent or experience. Some schemes, such as the Pompidou Centre, were proposing steel castings of unprecedented size and using steelwork without the usual protection from fire; others, such as the Dyce tent, were exploring the new world of prestressed membrane structures. The scheme for the Garden Festival building at Mannheim, was proposing a lattice-shell of timber laths, of unprecedented span and irregular doubly-curved form.

With hindsight, we know that all these proposals could be achieved, just as we know that the voyage Columbus was tackling was possible, that wafer-thin stone vaults in the early Gothic cathedrals could span 60 or more feet between perilously slender columns, and that Benjamin Baker's rail bridge across the Forth estuary was within man's capability at the time. But hindsight masks utterly the mental state of those who were involved in these projects *before* their respective solutions had been proven.

The 1960s and 1970s were a curious era in structural engineering history. Computers had arrived and were being hailed by many as destined to remove all the problems and uncertainties in analysis which were felt to be due to insufficient accuracy in calculations and the inability to solve by hand calculation the complex mathematics associated with large and unusual structures. Some people were coming to regard the output from computer structural analysis programs as infallible and there are many stories among engineers of buildings that might well have fallen down because of uncritical belief in computer output. Fortunately most disasters were avoided by the intervention of older and wiser engineers who had the experience to challenge 'infallible' results.

Opposite, from top left: Aerial view of the Mannheim Bundesgartenschau; Internal view of the lattice shell. The lattice fixed into its form by tightening the lattice load, fixing the lattice at the edge and by additional reinforcement with a broad mesh wire cable; Aerial view. The lattice shell is covered with PVC coated fabric. This membrane has since been replaced; View of the lattice edge; View showing the connection of the 300mm steel columns to the lattice.

Above, top: View across the lake; Bottom: interior view.

If anything, the atmosphere that pervaded the industry at that time makes the achievements of Structures 3 even more remarkable. Among all their innovative work there were no significant failures. They clearly had a robust working practice even at the very limits of what was possible. A considerable portion of the credit for this must fall to Ted Happold. In addition to his growing experience of unusual structures it was also he who studied closely the actual process of design, both in the abstract and in other manufacturing industries. It was he who used this understanding to consciously manage the design process on many of the unusual projects and thereby reduce the possibility of failure.

At the very end of his time running Structures 3, just before he left Arups in 1976, Ted wrote an article with his two colleagues Michael Dickson and Ian Liddell which summarised what they had learnt about designing unusual structures. The article 'Design towards convergence: a discussion', was published in *Architectural Design* – not a periodical often read by engineers. Perhaps for this reason, and because engineers seldom read widely around their subject, the article is now largely unknown. Its importance warrants substantial extracts here.

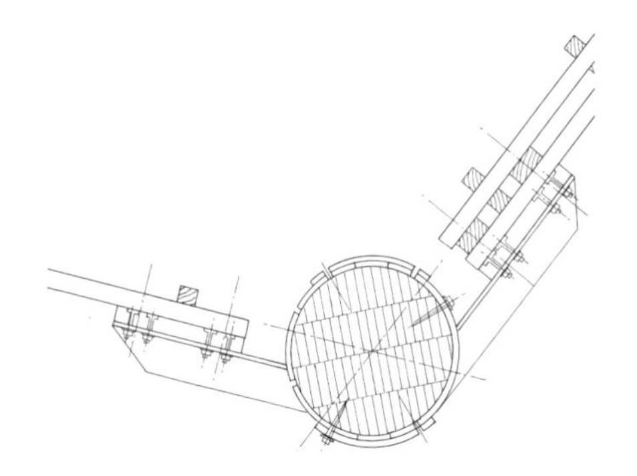

The article begins by summarising the widely accepted model of the design process as the cycle of analysis-synthesis-evaluation, repeated in increasing detail and converging on the final solution, one in which the broad outlines of the problems are tackled first, and then the subsidiary ones. The authors of the paper made a fundamental challenge to this model:

> We believe that the design process can be seen as a series of decisions which lead progressively towards the built reality. Each decision is made to satisfy the known requirements at that stage and to take into account those items of information which require further analysis before they can be fully defined. We believe that the process is a stage-by-stage one, and that the process in each stage can be described in the same terms – even though the content of each stage is different.

Four stages of the design process are identified – conception, form selection, detail design, and detailing for fabrication and erection.

> For the design process to be successful, it is necessary first to establish its objectives and significant order. An initial statement of intent helps to determine the relevant constraints and the balance between them. An answer in the form of a model can then be developed by the main contributor to each stage. The model has to be cast in such a form that it generates sympathetic models in the minds of the other contributors to the process and allows relevant design decisions to be made.
>
> The need for communicating the reasoning and philosophy behind the decision is paramount. If this cannot be communicated then it is more than likely that difficulties will arise in the future. The corollary is that the decision must be communicable.
>
> The evaluation of the performance model must be rigorous and is carried out by comparing it with the initial statement of intent. The concept of this Statement of Intent may develop as the design stage itself progresses. Really it is an initial listing of objectives and giving them values, followed by the development of their values during the modelling process.

Within each stage the Statement of Intent can be seen as a continuous control mechanism, looking ahead to possible future requirements or developments, and looking back at the objectives. It helps during the development of the models by providing continuous examination and, when used rigorously, provides an evaluation of the model at the end of the stage which, together with the collective experience and thoughts of the various contributors, forms the basis for the next stage. The theoretically cyclical design process is thereby changed into a linear, step-by-step one.

Formulating the Statement of Intent allows clear design objectives to be set and the creation, examination and development of appropriate performance models. These, in turn, allow communication of the design ideas and their evaluation. Conceiving and expressing the entire design process in terms of models which enable and facilitate communication within the design team was the vital key to making these abstract ideas useful in the design office and in helping to manage the design process.

The solution to a particular set of requirements is conceived as a model, initially a mental model. This model is developed and communicated in a variety of ways: description, drawings, physical models (representational or for form-finding), analogue models, mathematical or computer models. It is only by developing the model into a communicable form that the concept becomes imaginable and can be evaluated. As an example of this, the chain model of Mannheim is an excellent illustration of a model developed in terms of the architectural requirements of a second level of Statement of Intent. It gives a representation of the planning, the space enclosed, the structural form or an initial statement of the erection possibilities. Of course the end solution was not as simple as that, but as a communication tool at that stage of design it is excellent. This applies to the initial concept of a building, as well as to the concept of structural behaviour or the concept of an erection system. When used well, these models can provide the language that is required to create understanding at all levels. When the understanding is complete, the plan or structure becomes 'simple', and numerical calculation can often be based on easy formulae.

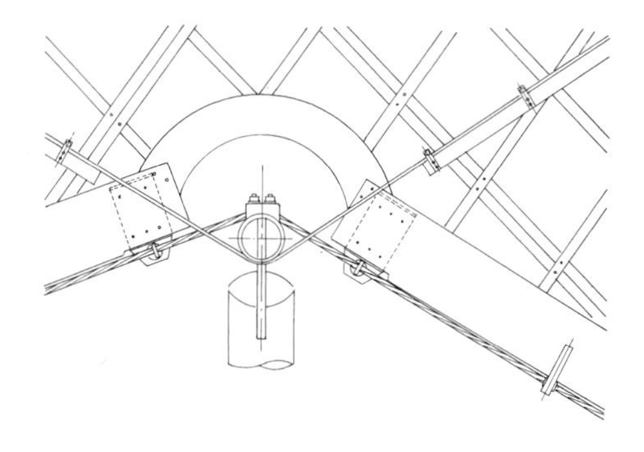

The four stages of design and Statements of Intent were illustrated by the Conference Centre in Mecca, the Dyce tent and, as summarised below, the exhibition hall and restaurant building for the Mannheim Bundesgartenschau with architects Carlfried Mutschler & Partners and specialist input from Frei Otto. This account of engineering design is rare in coming from the pens of designers – little of what is published about design is written by designers and even less by engineers. But it is also rare in its philosophical yet highly practical intent, conveying both details of particular projects and generalities relevant to all projects, and hence useful in informing the means by which the design process might be managed and controlled.

Conception

To start purposefully on a design the client's expectations must be clarified and perhaps this can best be done during the initial period by consciously expressing the overall restraints and objectives in preferential order. This effectively is the formulation of the upper level Statement of Intent where really the client is the main contributor. …

The garden show [at Mannheim] required the creation of a high-grade recreation park in which exhibition buildings of suitable form and of a temporary nature (8–10 year design life) were to be located. The garden

Opposite: Interior view of lattice shell – brought into its three-dimensional shape by the use of scaffolding towers; Drawing shows the Valley beam.

Above: external view during construction without the roof skin. Below: drawing showing cable boundary.

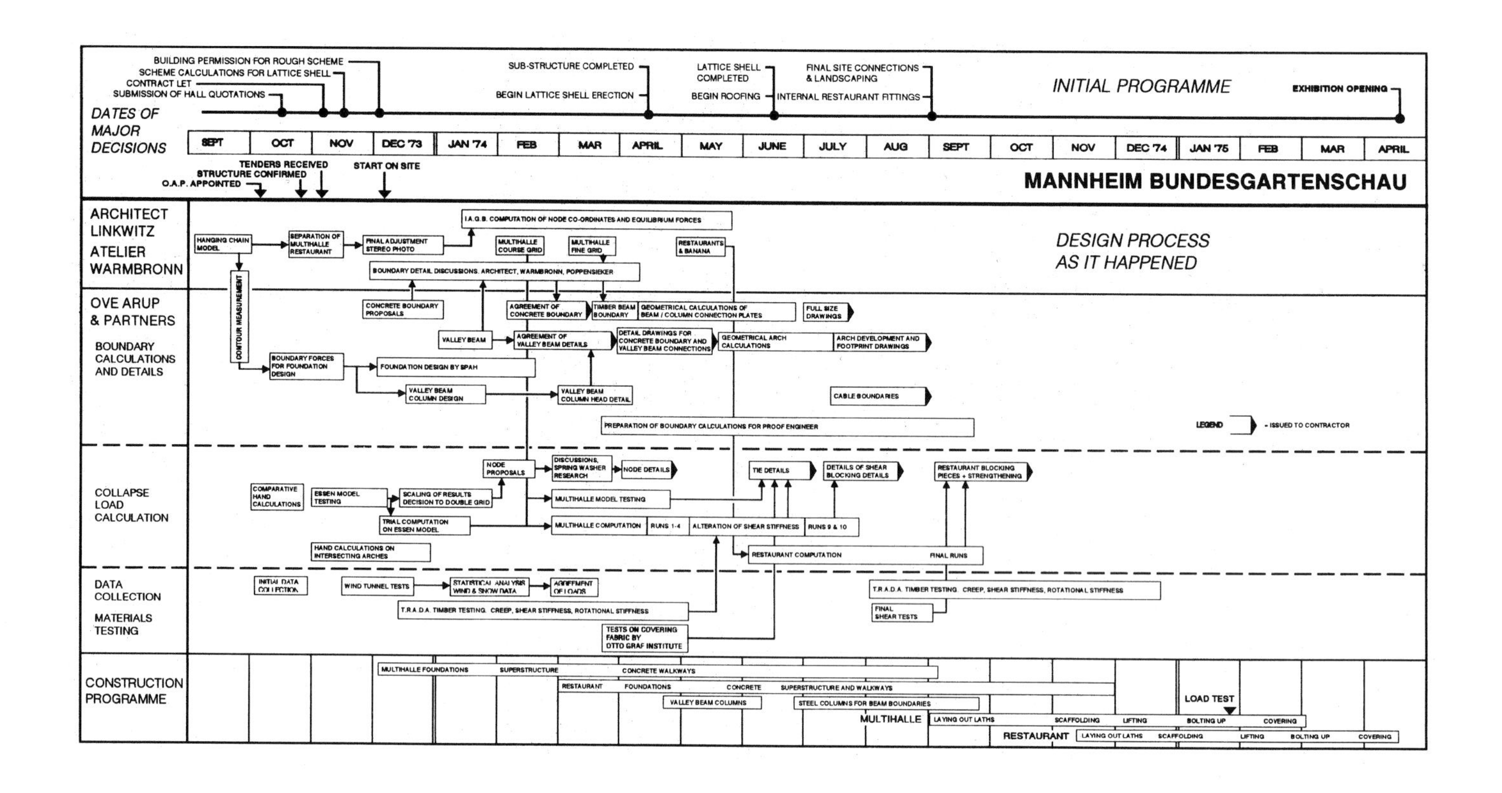
BUILDING PERMISSION FOR ROUGH SCHEME
SCHEME CALCULATIONS FOR LATTICE SHELL
CONTRACT LET
SUBMISSION OF HALL QUOTATIONS
SUB-STRUCTURE COMPLETED
BEGIN LATTICE SHELL ERECTION
LATTICE SHELL COMPLETED
BEGIN ROOFING
FINAL SITE CONNECTIONS & LANDSCAPING
INTERNAL RESTAURANT FITTINGS
INITIAL PROGRAMME
EXHIBITION OPENING
DATES OF MAJOR DECISIONS
SEPT
OCT
NOV
DEC '73
JAN '74
FEB
MAR
APRIL
MAY
JUNE
JULY
AUG
SEPT
OCT
NOV
DEC '74
JAN '75
FEB
MAR
APRIL
TENDERS RECEIVED
STRUCTURE CONFIRMED
O.A.P. APPOINTED
START ON SITE
MANNHEIM BUNDESGARTENSCHAU
ARCHITECT LINKWITZ ATELIER WARMBRONN
I.A.G.B. COMPUTATION OF NODE CO-ORDINATES AND EQUILIBRIUM FORCES
HANGING CHAIN MODEL
SEPARATION OF MULTIHALLE RESTAURANT
FINAL ADJUSTMENT STEREO PHOTO
MULTIHALLE COURSE GRID
MULTIHALLE FINE GRID
RESTAURANTS & BANANA
DESIGN PROCESS AS IT HAPPENED
BOUNDARY DETAIL DISCUSSIONS. ARCHITECT, WARMBRONN, POPPENSIEKER
CONTOUR MEASUREMENT
OVE ARUP & PARTNERS
BOUNDARY CALCULATIONS AND DETAILS
CONCRETE BOUNDARY PROPOSALS
AGREEMENT OF CONCRETE BOUNDARY
TIMBER BEAM BOUNDARY
GEOMETRICAL CALCULATIONS OF BEAM / COLUMN CONNECTION PLATES
FULL SIZE DRAWINGS
VALLEY BEAM
AGREEMENT OF VALLEY BEAM DETAILS
DETAIL DRAWINGS FOR CONCRETE BOUNDARY AND VALLEY BEAM CONNECTIONS
GEOMETRICAL ARCH CALCULATIONS
ARCH DEVELOPMENT AND FOOTPRINT DRAWINGS
BOUNDARY FORCES FOR FOUNDATION DESIGN
FOUNDATION DESIGN BY SPAH
VALLEY BEAM COLUMN DESIGN
VALLEY BEAM COLUMN HEAD DETAIL
CABLE BOUNDARIES
PREPARATION OF BOUNDARY CALCULATIONS FOR PROOF ENGINEER
LEGEND - ISSUED TO CONTRACTOR
COLLAPSE LOAD CALCULATION
NODE PROPOSALS
DISCUSSIONS, SPRING WASHER RESEARCH
NODE DETAILS
TIE DETAILS
DETAILS OF SHEAR BLOCKING DETAILS
RESTAURANT BLOCKING PIECES + STRENGTHENING
COMPARATIVE HAND CALCULATIONS
ESSEN MODEL TESTING
SCALING OF RESULTS DECISION TO DOUBLE GRID
MULTIHALLE MODEL TESTING
TRIAL COMPUTATION ON ESSEN MODEL
MULTIHALLE COMPUTATION
RUNS 1-4
ALTERATION OF SHEAR STIFFNESS
RUNS 9 & 10
HAND CALCULATIONS ON INTERSECTING ARCHES
RESTAURANT COMPUTATION
FINAL RUNS
DATA COLLECTION
MATERIALS TESTING
INITIAL DATA COLLECTION
WIND TUNNEL TESTS
STATISTICAL ANALYSIS WIND & SNOW DATA
AGREEMENT OF LOADS
T.R.A.D.A. TIMBER TESTING. CREEP, SHEAR STIFFNESS, ROTATIONAL STIFFNESS
FINAL SHEAR TESTS
TESTS ON COVERING FABRIC BY OTTO GRAF INSTITUTE
CONSTRUCTION PROGRAMME
MULTIHALLE FOUNDATIONS
SUPERSTRUCTURE
CONCRETE WALKWAYS
RESTAURANT
FOUNDATIONS
CONCRETE
SUPERSTRUCTURE AND WALKWAYS
VALLEY BEAM COLUMNS
STEEL COLUMNS FOR BEAM BOUNDARIES
LOAD TEST
MULTIHALLE
LAYING OUT LATHS
SCAFFOLDING
LIFTING
BOLTING UP
COVERING

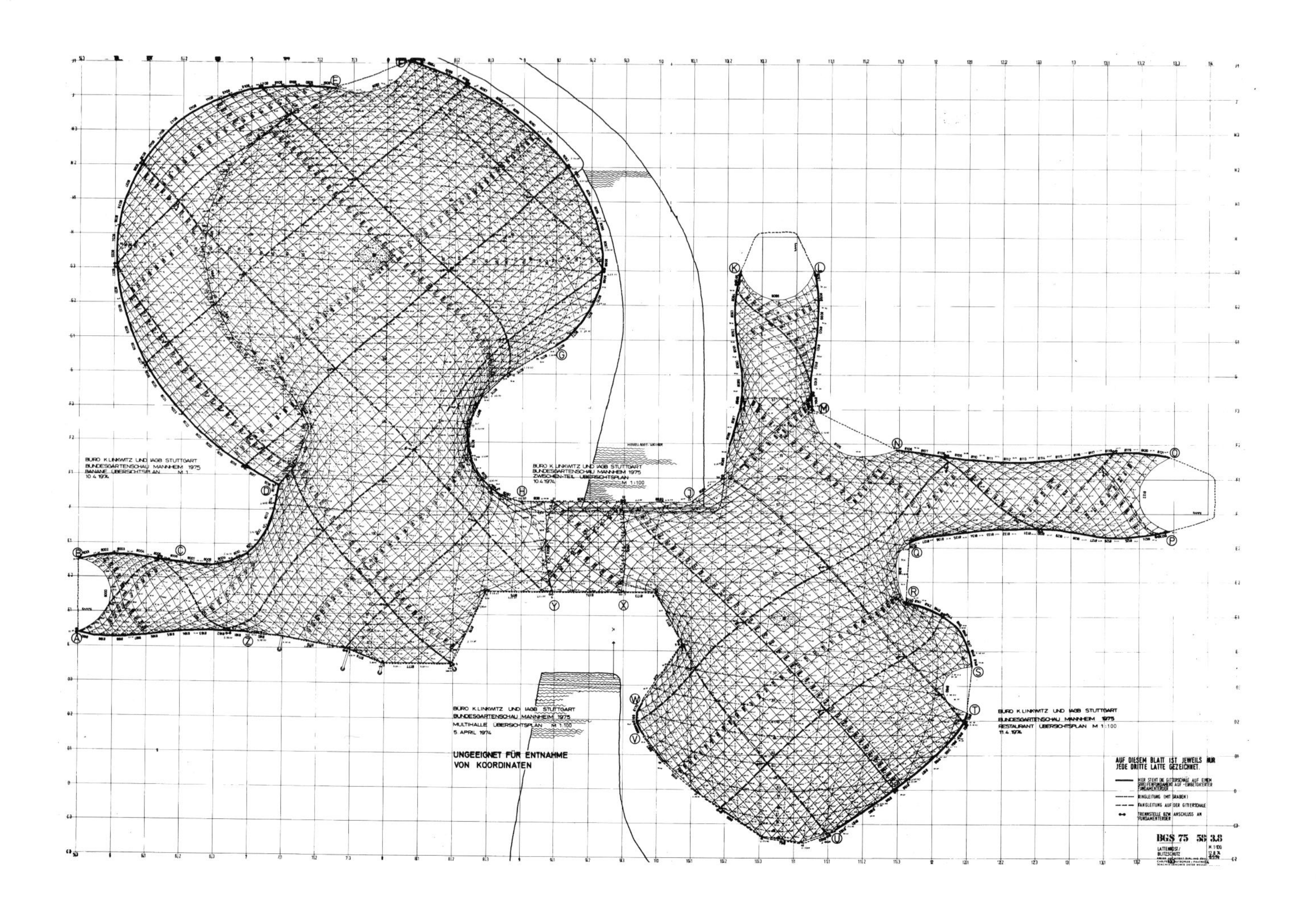
UNGEEIGNET FÜR ENTNAHME VON KOORDINATEN
BGS 75

show is a paying exhibition which is sponsored by the government, so that, apart from political pressures, cost was not of prime importance. However construction time was pressing – the show was to open in 18 months. The proposal for the roof over the Multihall was a tent, supported by balloons.

Selecting a form:

The second layer of design is principally architectural, but the client must understand it and the engineer be willing to underwrite it. This level starts by identifying the objectives and limits of the project as a whole and entails the formulation of a second level of Statement of Intent through which the development of the concept is controlled.

An essential part of formulating intentions and approach to the second layer of design is a Time Plan, because choice is affected by time, and constant monitoring against a time plan is useful. The relative merits of various built forms can now truly be investigated.

In long span structures, we believe that primitive physical form models are helpful, almost indispensable to establishing an order to the overall design. They enable the engineer to compare design alternatives by the use of simple mathematical models to which well-known analytical formulae, coupled to mental analogue models, can be applied. The engineer can then explain in broad terms to the architect the choices available and, in turn, perceive and discuss the qualities desired by the architect. The client can also understand the broad decisions. More sophisticated physical or mathematical modelling at this stage is not required. …

In our practice [Arup's at that time], the engineer is encouraged to build up his own overall mental model which is quasi-numerical, and prepare its defence against other views. This helps communication to different parts of the team. From the application of numerical analysis to models of possible forms and construction materials, both in terms of specific engineering characteristics, and in terms of the levels of approximation involved, structural comprehension gradually emerges. …

Closer observation of the Mannheim competition scheme, in the light of the building regulations, and when viewed as an internal external building in a park landscape, prompted Professor Otto to suggest an alternative idea: a single, continuously curving, free-form grid shell, instead of the original two separate halls supported on helium balloons. A primitive wire model was made of this idea and compared with the desired objectives. This wire model was developed by Prof. Otto into the 1:100 hanging chain model which, in fact, defined the geometry of the final roof.

Detail design:

The convergence of the design comes out of the cross-evaluation of many different types of model, each involving different members of the engineering team. Wind tunnel testing of aeroelastic models may be used to test stability under wind, or flexible structural models may be used to simulate collapse behaviour. The final numerical assessment of strength will be made using complex mathematical models in a computer, but coarse and more easily understood models may be used to help create confidence in the computer results. …

The detail design of Mannheim was unusual from two points of view. First, the engineers from Ove Arup & Partners were summoned to the scene only some three weeks before tenders were due back, and about 16 months before the show was due to open, after the existing engineers had

Opposite: Flow chart of the design and construction programme for the Mannheim Bundesgartenschau; General arrangement drawing of the structure.

Details of the grid were developed in parallel with the competition work so that the properties of the structure agreed with those of the mathematical model which was finally accepted as having an adequate factor of safety. The boundary details were developed to meet the requirements of the geometry and the erection process and to have adequate strength to supply the reaction to the lath forces.

Above: Load experiments were carried out to establish the load-bearing capacity of the lattice shells. Dustbins filled with water were hung at every ninth node and the deformation measured by suspended plumb-lines.

resigned their commission. At that stage, while the form was almost entirely decided and the materials of construction agreed upon, the structural design was still in its infancy. The role of the roof in the total concept had been established, but without the engineers underwriting the preliminary studies by evaluating possible problems and avoiding those that appeared likely to be difficult by choice of design.

Secondly, small grid shells had been built before but not at this scale, and not ones requiring careful consideration of shell buckling and large displacements. There existed then little stored knowledge for this type of structure. The immediate need was to underwrite the basic structural requirement for the two 30 x 55 and 60 x 80m large shells. Based on simple physical model analogues and two-way arch analogues, sufficient understanding of the likely overall behaviour was built up.

It was decided that a double layer of laths would provide the basis for the structural solution. Additional simple calculations enabled evaluation of boundary types and foundation loads which were required to install the foundations. A perspex strip model was then made to simulate buckling behaviour. This led to an improved understanding of the overall behaviour. Material and member testing indicated that this model did not adequately represent the behaviour of the real structure. A mathematical model was set up, using a non-linear program, which finally provided numerical values of the collapse load, and indicated areas of the shell where extra stiffness was required.

Detailing, fabrication and erection:

For the fabrication of individual components there is often a problem with complex geometry. This is now eased by the progress of computer technology, but there is still the need of the fabricator to understand the geometry. This is best achieved if he does the shop drawings. A more serious problem remains, that of choice and responsibility for the construction method.

Communication from designer to contractor of confidence, structural understanding and the need for accuracy and control is often limited. Simple two-dimensional drawings are inadequate as the sole means of modelling during the design – so why should tender drawings and documents be adequate for the contractor? For long-span structures, particularly special one-off ventures, the choice of the right construction method is absolutely vital if the potential economies are to be realised. This, in turn, requires a precise understanding of the nature of the structure to be erected, and the methods by which this can be achieved.

Who is better than the designer for this role? In theory, no-one, since presumably the constraints of construction are thoroughly considered in the development of his design. In reality, however, he does not control the men and machines on the building site, nor the expenditure on these items. Therefore, at the time of design, he must be somewhat ignorant of the methods to be adopted. ... Our grim conclusion to date has been that, for special ventures, some additional means of communication to tender documents is required. ... One aid to communication has been the use of the 'trial structure'. This hastens the learning curve and concentrates the faculties on the essential erection sequences. The whole responsibility for the conduct of the trial is set with the contractor. Also it clears the division of responsibility between fabricator and erector – or rather between on-site and off-site work. Both constructor and designer observe the process. Explanations and discussions between designer and contactor are stimulated by the observed physical characteristics, enabling firm alternatives to be selected. The choice of the scale of initial structure is important. it can range from full size (i.e. a part of the final building) to a 1/4 size replica or to a small scale model. In all of these the prime engineering characteristics of the structure during erection must be scaled. The exercise must be a true and understandable analogue model which can convey confidence and understanding in the chosen system from designer to contractor. ...

The engineers [on the timber gridshell at Mannheim] had assumed that the contractor would prepare the fabrication drawings and had excluded these from their work. It turned out that the contractors had also excluded this part, claiming that they would only work to the engineers' instructions. To break the deadlock, the engineers undertook to do all the geometrical drawings for the boundary members and their connections. In some ways this was beneficial, as the engineers were able to iron out inconsistencies in the computer co-ordinates, but the job on site suffered because the contactor had not been made to study the geometry. With the drawings being made in London, the timber fabrication being carried out in northern Germany, and the steel fabrication being made in Mannheim, there were several examples of members which wouldn't fit.

The engineers were appointed after tender documents had been sent out and these had suggested the use of cranes to lift the lattice. The analytical work suggested that it was necessary to use a double layer lattice and this meant that the shell would not be much floppier during lifting and would not be stable until all the bolts were fixed.

The disadvantages of using cranes were pointed out to the contractors at quite an early stage, but they were not prepared to listen. They were afraid of losing money if the method was changed; in any case they did not believe that the shell would stand. Time did not permit a trial structure, so to demonstrate to the contractor the problems of using cranes, a weighted wire-mesh model was made which scaled the deformations of the unbolted lattice. This model finally convinced the contractor that cranes were impossible, and it was proposed that scaffold towers should be used. Only the removal of the towers and the load test finally convinced the contractor that the shell would stand.

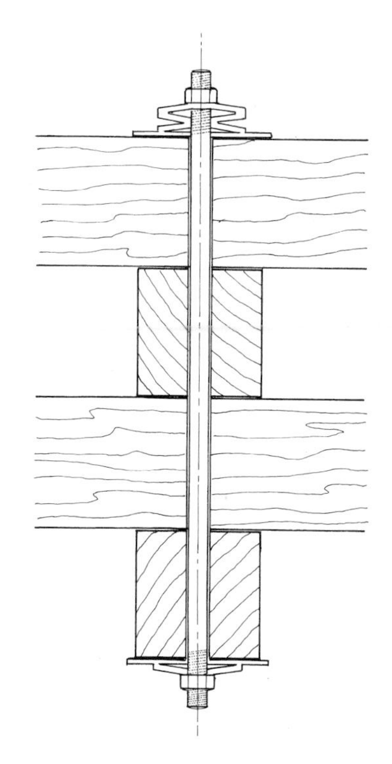

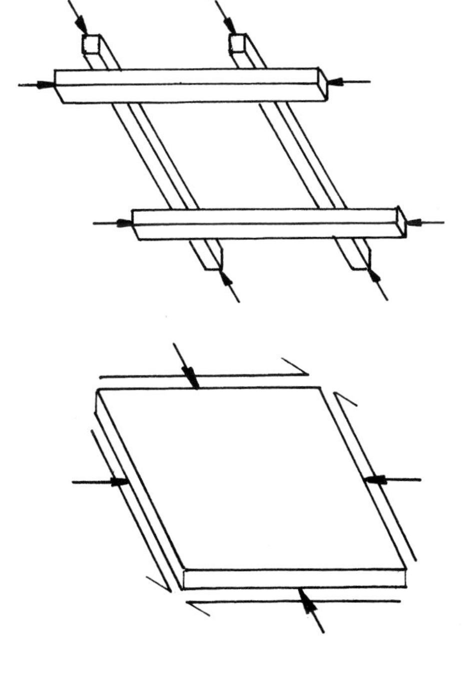

As an account of engineering design this is a lot less dry than most that appear in the technical press but neither in this nor the several other published accounts of the design of this building is the sheer brilliance and ingenuity of some of the details conveyed. The entire success of the timber lattice depended on the node joint between each pair of 50 x 50mm laths made from western hemlock. The connection had to transmit shear forces between the two layers of the lattice to give it in-plane stiffness. However, it also had to allow considerable shear during erection to allow the orthogonal lattice to deform into diamonds of varying angle to create the required curvature in the third dimension. The joint could not, therefore, be a mechanical, rigid connection; it would have to carry the shear forces by means of friction between the laths. This could be generated easily enough with a bolted connection and spring washer, but this would slacken with time as the timber dried out and shrank. What was needed was a spring washer that resists with constant force, independent of the degree of compression – a clear contravention of Hooke's law of elasticity. The solution was to use disc springs which, as they are compressed towards being flat, by virtue of their conical form, do indeed approach the flat load extension behaviour required – closely enough to make this connection work, anyway. And that, as Ted said, is engineering.

Opposite, top: Lattice shell under construction; View of the interior lattice.

Above. top: typical node joint below lattice shell and continuous shell elements

The Politics of Engineering

Ted Happold seemed on occasion to be a throwback to the heroic generation of men engaged in the massive constructional operations which transformed the landscape and environment of nineteenth-century society. Angus Buchanan, writing about the lifestyle of British engineers, felt the most striking characteristic about the work of the nineteenth-century engineers in Britain was its quantity: as a group and as individuals they tended to work exceedingly hard. Conder records railway engineers working for days on end in order to meet the Parliamentary timetable for submitting schemes for legislative approval, and there is plenty of reason to think this was a regular and even habitual occurrence. I K Brunel drove himself and his staff through long periods of almost unremitting labour, and other engineers like Robert Stephenson, C B Vignoles and G P Bidder behaved in the same way. The communal character of engineering does not appear to have carried over into camaraderie beyond the workplace and the embryo professional institutions. The fact of the matter seems to be that the Victorian engineers generally did not care much for socialising. They rarely took any part in political activities, and when they did it was not with much distinction. Both Locke and Robert Stephenson became Members of Parliament, but were as undistinguished in that milieu as they were pre-eminent in their profession. I suppose then the politics of engineering should be defined with a small 'p' – more about networking, communicating, establishing patronage, influencing management within the construction sector, oiling the wheels of establishment institutions, focusing on the areas of maximum opportunity and establishing trading programmes and information quality to keep ahead of the pack.

When I spoke with Michael Dickson and Graham Watts about Ted's extraordinary range of activity outside the practice, academia and his family, we were all unanimous that his will to succeed was always tempered with a strong sense of public concern. As a child around the family table he was made aware by politically active parents that the greater good of the community was important and that part of his life must be spent helping others less fortunate than himself, and that it was his duty to strengthen and support those bodies and institutions that were dedicated to providing better opportunities, better workplaces and better education for the public at large.

The strong Quaker beliefs of his family and their closest family friend (Ted's mentor and role model T Edmund Harvey) were deeply ingrained. He absorbed the best qualities of his native county – loyalty, good humour and an extraordinary work ethic. He was, however, at first a little insular for the London scene. Povl Ahm on first encountering Ted said 'All he could talk about was his parents and after his period in New York the subject matter increased to his parents and America'. But that introspective period did not last long. Michael Barclay, Ted's contemporary at Arups, recalls that 'Ted was never like anyone else. In the first place he was never in the office and secondly he seemed to have no staff, just working on special projects with Povl Ahm.'

Coming to Arups in 1963 from the contracting side of industry, Barclay entered a totally different world, full of disorder, indiscipline and incredible inefficiency but possessing a sense of freedom and creativity in which anything could happen. This was exactly the fertile soil Ted needed to thrive in. Ted became one of Arups group engineers and in that capacity the entrepreneurial side of his nature blossomed. Strange rumours circulated around Arups from Ted's group … he was holding lunchtime meetings on the theory of design, he was active in the Junior Liaison Organisation (JLO), he even had architects staying with him at home. There was no one else in Arups outside the magic circle of top brass who was in that class … and thus was born the Happold maxim for success: 'You must have style!' – and he did have style. He became Chairman of the Junior Liaison Organisation. Young professionals gathered from all the chartered institutions, which gave impetus to his ceaseless networking and his ability to influence and advance visionary ideas in both his professional activities and his hopes for the construction industry as a whole.

After his success in Saudi Arabia in negotiating two of Arups most profitable contracts for the Conference Centres in Riyadh and Mecca, he arrived back to London to find, in his own words, (his biggest failure) awaiting him. Basil Spence's Cavalry Barracks in Knightsbridge had been allowed to go on site with incomplete drawings. Sir Robert McAlpine had stopped the job after two weeks and put in a claim for damages. They had substantiated their claim with apparently indisputable figures based on their new critical path management programme – a terrible weapon in those days. A lot was at stake for Arups and for Ted personally. Though ill after his Saudi efforts he came back with a rush, banged heads together, organised a rescue operation with the Spence offices, limited the damages by a combination of hard bargaining and political skills; he then cleared his books with Arups and put the whole thing behind him. This combination of skill and hard-nosed construction leadership demonstrated to Michael Barclay, whose group was being merged with Ted's at that time, that the senior partners of Ove Arup, after much horse-trading and debate, were right in appointing Ted to lead their third structural division with Peter Rice as Associate Analyst and himself as Associate Project Manager. Without asking for permission, which would probably have been refused, Ted renamed the division Structures 3 and began to let the world know about it. It was reminiscent of Winston Churchill's return to the Admiralty in 1939 … he took over with relief and joy knowing that at last he was where he wanted to be. With the success of Structures 3 he established a unique platform for his wide ranging ambitions. His workload was prodigious and with regular competition wins and a truly international cast of architectural collaborators he was able to pursue a vision for his future which saw him combining the academic and research role implicit in his genes with high profile practice and a wider role as an industry spokesman which he felt was his duty.

In 1976 Ted left Arups to take up the Chair of Building Engineering at the University of Bath – and simultaneously started Buro Happold. The move was to cause an explosion within the Arup organisation – his was the first and only defection from the leadership. It was, I think, a singularly unpleasant experience for both parties; both felt they had grievances and both felt a sense of personal betrayal. Ted had been with Arups for eighteen years, and had contributed a great deal to one of the world's outstanding engineering practices, both in making Structures 3 one of the most formidable design groups in European engineering, organising with Poul Beckman the Arup in-house training programme and building up a high profile and profitable body of work for the company. He had in turn had the space and time to develop the experience and skills honed with Arups' support and backing, which made him a formidable candidate for both academia and private practice. The move

Opposite, top: I K Brunel standing in front of the chains of the SS Great Eastern, c1857; Bottom: Joseph Locke. Above: Robert Stephenson.

to Bath and the formation of Buro Happold. Ted made it clear in his Statement of Practice Intent that 'Buro Happold is a professional consulting engineering practice. A profession is defined as having a body of knowledge, possessing diagnostic and problem solving techniques and holding an attitude to service which is not just about money but working for the public good to a code of professional conduct. Consulting means we are for hire. Engineering takes scientific and traditional knowledge of the physical and human environment, together with an understanding of construction methods and the market to join clients, architects, contractors and others in providing solutions to problems. It is about economy and value.' Buro Happold is owned by a partnership set up in a fellowship before it became an organisation. A partnership because the founders thought their skills were complimentary, and that the whole would be greater than the parts. The fellowship is the core of it though, and represents not only mutual confidence and forbearance but similar standards of business conduct.

The political side of Happold's nature was based firmly on three scenarios. The first was to benefit his practice by utilising his talents, his time, his persuasive skills and virtuosity to place himself in as many academic, business and institutional situations in order to widen his network of contacts. The second was to polish a healthy ego and perhaps respond to a lifetime's visionary search for total recognition for himself, his partnership and his profession. He had tasted success at a very early age and felt, like Ove Arup before him, that success only continues if you maintain standards, pursue each opportunity with drive and supreme marketing skills and surround yourself with focused and talented colleagues. The third was more intangible. All his writing, many of his lectures and much of his conversations were imbued with a social agenda. He was dedicated to a type of public service that would strengthen his chosen industry – Construction, which he felt had been hopelessly fragmented in the twentieth century – and the institutions to which he belonged … that would bridge the gap between science and the arts in the richest sense of the word and, much more importantly, produce a climate of constructive intellectual creativity which would benefit society as a whole. A very tall order for a dedicated family man confronted like all of us with only twenty-four hours in a day.

The little spider's web diagram on the green inner leaf shows how the political strategy evolved from modest beginnings at the JLO to the real leadership role he played in the construction industry in the last decade of his life. President of the Institute of Structural Engineers, first Chairman and co-founder of the Construction Industry Council, a member of the Building Regulations Advisory Board, Director of the Property Services Agency Advisory Board. He was also tying the knot between science and the arts – first with his appointment as a Royal Designer for Industry, a member of the Design Council, Vice-president to the Royal Society of Arts and, in 1992, Master of the Royal Designers for Industry. His committee activities were equally important and extraordinary in variety and density of application. He chaired the committee in the Institution of Structural Engineers that produced the Appraisal of Existing Structures in 1982 and its second coming in 1990. He was a member of the Popplewell enquiry into the fire at Bradford football stadium. Following this he formed the Institute of Structural Engineers Working Party on the appraisal of sports grounds. As Hugh Johnston observed of Ted's mastery of the committee format: 'It was not just the inspiration, insight and enjoyment that he invariably brought into design discussions that was so valuable, but also his astute political sense and tenacious diplomacy in much wider policy and organisational debates.'

When it came to founding the Building (later, Construction) Industry Council Ted needed all these qualities and more in order to bring together the essential partners and achieve not just the official launch but positive forwards so that the remaining doubters could be convinced and initial scepticism countered. Ted himself always said that like all construction projects, setting up the CIC was a team effort and indeed it could never have been achieved on any other basis. The fact remains, however, that it was Ted's vision and initiative which provided the first spark and his leadership, good humour and freely given time were crucial to its success. I have a copy of the proposal to found a Building Industry Council and the messianic zeal of Happold is encapsulated in *Why?*. The building industry only exists to serve society and the industry's overall objective should be to serve and protect the consumer. To achieve this there is a need for collective decision-making and, therefore, unity. We, architects, builders, services engineers and structural engineers, propose to found a Building Industry Council. We are from the Chartered Institutions already listed and it is hoped as well as expected that those who share these ideals will soon join this organisation. Building is a uniquely complex group activity. Even within the building industry there are many problems, partly because there are public misconceptions about individual contributions. The results are that society often does not know what abilities the different groups in the industry have, and are either vulnerable to 'cowboys' or disappointed by the service. Recruitment for vital sectors is affected."

The CIC proposes to promote and encourage export of the services of the industry ... determine better ways of minimising the costs of latent defects, ensure fairer competition for design contracts, encourage and assist the building materials industry, advise and assist in the provision, content and standards of training for all skills, and assist nationally and internationally in producing building regulation standards and codes ... consider registration of professionals in the building industry and ensure compatibility and quality ... and last but not least develop, encourage and co-ordinate as necessary the industry's research programme.

The CIC's famous charter is signed by the, President, Vice-President and immediate past President of all four institutions: the Royal Institute of British Architects, the Chartered Institute of Building, the Chartered Institute of Building Services Engineering and the Institution of Structural Engineers. Happold used his 'speakers' corner' journalistic activities for the *Builder* to nurture, explain and propagate the CIC ... an institution which achieved what many people thought was impossible – all the professional institutions sitting at the same table and reaching agreement on matters of common interest to their 300,000 members.

Insular Britain was never a large enough playground for Ted Happold and as his research and project interests stretched to each continent his public platform shifted to the International Association of Bridges and Structural Engineers based in Switzerland and the International Association for Shell and Spatial Structures based in Madrid. At their conferences and in their working groups he was capable of playing the British eccentric in almost any company, and it is significant that his much valued contributions were celebrated at the IASS Conference in 1996 which was dedicated to his memory. The spider's web was truly international and if intergalactic travel had been available one would have no doubt who would have been the first engineer on board.

Opposite, top: Contemporary cartoon in *Punch* of Sir Joseph Bazalgette, the doyen of public health engineering; Bottom: Robert Stephenson in committee. Celebration of the Britannia Bridge contemporary painting, c1850.

Above, top: Sir Joseph Paxton's original sketch showing the germs of the crystal palace c1850; Crystal Palace 1851 Exhibition Committee in session.

Design Education: On Being and Becoming an Engineer

Most who have become university academics did so out of a passion for research. This was not the case for Ted Happold. He took up his post in 1976 as the Professor of Building Engineering at the University of Bath for three main reasons. After several years of the remarkable success of Structures 3 it had become clear to Ted that he might not find his way to the very top at Arups and the move to a university was one future avenue he set about exploring. As the son of a professor of biochemistry he had grown up in a university climate and, despite Ted's success as an engineer, his father had never felt that this was quite as laudable as being a successful academic. Ted admitted in discussions with many of his friends at the time that, deep down, this was, perhaps, a greater influence than he realised. But the third reason was probably the dominant one – he saw a professorial chair as the opportunity to try to change the way building engineers were educated and to establish research programmes of direct concern with engineering design.

Ted's interest in educating young engineers had begun in Arups ten years earlier when the practice was expanding and taking on increasing numbers of fresh graduates. It was clear that they needed help in approaching projects with the right sort of attitude and basic skills and he proposed starting an Arup Graduate Training School. The idea was developed by Ted with Poul Beckman and it ran for the first time in 1966. Trainees attended until Alan Frampton and David Cathro, who ran it day-to-day, felt they were ready to leave – usually after about ten weeks. It taught the basic communication skills that engineers need – printing, drawing, setting out calculations and (in those days) doing reinforcement detailing. The final test was producing a scheme and calculations for a project in a simulated real world – where architects and clients change their minds half-way through – and then passing the work on to someone else to detail the reinforcement. To this rigour Ted added his own passion for design methodology. This included brainstorming as a problem-solving technique for generating ideas and refining them into a practicable working scheme. How, for instance, would one get across a sea of treacle? He used this method in his own design group too. On one occasion he involved everyone in the office, secretaries included, in developing ideas for a competition to design new covers for the cricket pitch at Lord's. They won the competition but the project went no further.

Ted arrived at Bath in typical style. It was said that he almost made it a condition of accepting the Chair of Building Engineering that he be allowed to change the structure of the undergraduate courses so as to have students of different design disciplines – architecture, structural and services engineering – studying together, and with large parts of the courses in common. He believed passionately that young people who were studying to become designers in the construction industry should undertake design projects with those intending to work in other sectors of the same industry. While at Arups he had often tutored student architects at the University of Cambridge and seen at first hand the results of their ignorance of building engineering. At Bath he achieved, by and large, the integrated approach to the education of engineers and architects he sought and his fundamental ideas still underlie the undergraduate courses there. It is worth reflecting, however, that it took a man of Ted's energy and

A servant of good architecture

At the invitation of the Council, Mr. Edmund Happold (F) prepared a paper 'The Household Cavalry Barracks, Knightsbridge' for presentation at the Public Works Congress in November 1970 in the session sponsored by the Institution. The text of the paper and of the discussion among the small audience that followed its presentation appears in the *Conference Proceedings* which is shortly to be published.

In introducing his paper, Mr. Happold referred to the role of the structural engineer as designer and to his relationship with the architect. The Council feels that this expression of Mr. Happold's views is worthy of a wider audience.

The Institution of Structural Engineers asked me to prepare a paper on the rebuilding of the Hyde Park Cavalry Barracks for presentation at the Public Works Congress 1970, because, it was thought, it would illustrate what the structural engineer contributes in a building project. I at first thought that this was a mistake; if we wanted to show what an exciting contribution the structural engineer can make to a project why not produce a paper on an example where the structural engineering aspects are dominant? (Figs 1 and 2.)

Fig 1.

Then I thought further—why do I stay in architectural engineering and not move back into the field where the structure is the major part of the total work, where the structural engineer is king and compromises, as I see them, do not have to be made? Where the architect is under the engineer's control and cannot wear the engineer down. Where the status and admiration comes to the engineer as chief designer. Where there is not that grubby image of the building industry—an image the industry has had for a long time—always late delivery, always over the cost limits.

I know the answer to that. I work in interdisciplinary teams because I believe that different professional skills working together usually reflect different talents. These different talents, when they are talents, challenge and educate me and can stimulate me to contribute to producing a better job—maybe make me continue to produce better and better work.

But what do I mean by a better job—what is my aim as an engineer? At one level it must be to do good engineering. To design the structure which is the cheapest to build while fulfilling all the functions it is required to perform. Functions such as an ability to stand up and carry all the loads which the client tells us will be imposed and which nature imposes on the structure; ability to last, with minimum maintenance, the length of time the building has to last; ability to expand and contract without cracking as temperatures change; not to deflect

Edmund Happold, BSc(Eng), CEng, FIStructE, FICE, is the Executive Partner of Structures 3, one of the structural engineering divisions of Ove Arup & Partners, consulting engineers. Educated at Leeds Grammar School, Bootham School York and Leeds University, he has worked for both contracting and consulting firms in England and the United States.

THE STRUCTURAL ENGINEER JULY 1971 No 7 VOLUME 49

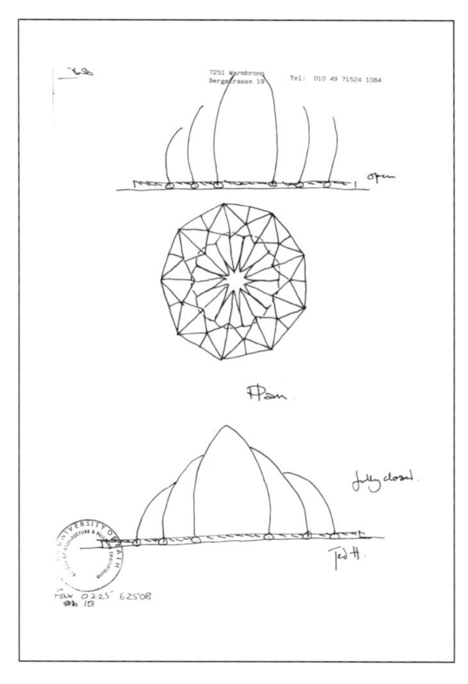

purpose to achieve this; there was resistance from both inside and outside the department. While his ideas about bringing architects, structural and services engineers together in Bath was accepted, or at least tolerated, some colleagues felt he had really gone too far in bringing construction management into the fold. This he nevertheless achieved in 1990 in typical manner: he discovered that the construction management staff at Brunel University were finding their position in the engineering department there increasingly difficult to sustain and Ted used his many and varied contacts in the industry and the university world to coax the entire group of staff to join him at Bath.

Ted's purpose in reshaping the courses at Bath was underpinned by his deeply felt ideas about the nature of engineering and the engineer's role in society. He had thought often and profoundly about what it was to be an engineer and wrote about it more frequently than anyone. His views came from reflecting on his own experience, of course, but were considerably extended by his wide reading in both the history of engineering and the philosophy of technology. He passed on his ideas to the students, especially in his ten annual lectures on the history of engineering which were among the best attended by the first-year students of all the disciplines and inspired more than one first-year architecture student to change to engineering! He felt strongly that all engineers should have a knowledge of the history of their profession and its key players, not out of idle interest, but as part of establishing 'the culture of our industry' and 'claiming our history'. He talked mainly of people, not of things – about their engineering, commercial and political skills and the stamina and sheer determination of people such as Telford, Brunel, the Stephensons and Fairbairn. He felt it was vital that students learn, and pass down the generations 'the thought processes of past-masters of our crafts and of the methods of our industry' and that they equip themselves with 'a lexicon of examples of design solutions'.

While driving home the need for engineers to understand the ways of architects, and vice versa, Ted certainly did not advocate architect-engineers or engineer-architects. He was very clear about the distinction between the two professions. Rather, he was keen that young engineers should develop a strong sense of their own identity – another reason why he insisted on the history lectures. He brought many of these ideas out in his address as President of the Institution of Structural Engineers in 1986. He entitled it 'Can you hear me at the back?', alluding to the fact that many structural engineers choose to take a subordinate role in the processes of designing and constructing buildings. He used the address to convey a rare appraisal of what good structural engineering is all about and, by implication, what more people in the industry should aspire to and achieve:

On the profession of structural engineer:

> We must question our profession's present role and standing – review how our skills and knowledge have altered over the years and try to determine those skills which society needs and which we think we could provide .... The need for this debate is unfortunately reinforced by the current vigorous commercial promotion policy by other professional groups who claim they can do what we do, hire us to do it, and then appropriate authorship. Such scenarios not only mislead the public but also lessen the profession's influence. The public's perception of engineering influences recruitment into our profession and could certainly affect the allocation of national resources for eduction, training and research. It is important that you all join in this debate because what we decide to do will define the future of our profession.

Opposite, top: Ted in teaching mode; Bottom: In an early polemic Happold takes the engineering party line! In later life he repudiated this notion and quite rightly supported a partnership of complementary skills between the architect and the engineer.

Above, from top: Finding form through modelling was a Happold staple. This party piece shows the practice working on the structure for Vail with the completed model below. Ted continued to work through drawing and debate throughout his life. Bottom: project with Matthew Priestman showing Ted's sketch for transmission to Frei Otto just before his death.

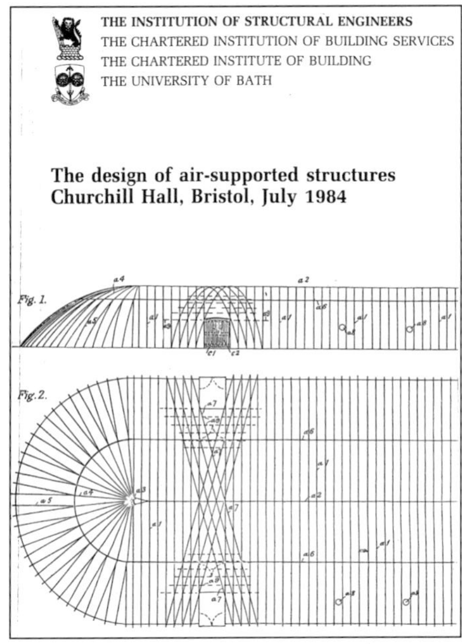

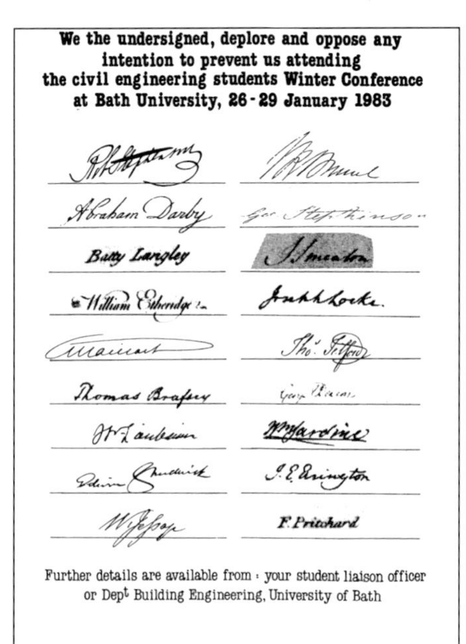

We the undersigned, deplore and oppose any intention to prevent us attending the civil engineering students Winter Conference at Bath University, 26-29 January 1983

Abraham Darby

Batty Langley

Thomas Brassey

F. Pritchard

Further details are available from : your student liaison officer or Dept. Building Engineering, University of Bath

On architect and engineer:

> Architects do not feel secure – a profession that is trying to sell 'value' without quantifying it is bound to be vulnerable – though it might be the right way to sell it. It is perhaps not surprising that some architects imply to clients that they not only have their own skills but the engineer's as well – surprisingly enough, helped by engineers who see themselves as dependent on the architects for work.

On education:

> To teach [engineering] students to state clearly to non-specialists what they do and why, is the test of our education. Our position in society and our skills are related – if we have declined in any area, it is in being an effective part of the decision-making process. If you look at the diaries of such as Jessop you will see how much of their time was spent convincing clients or Parliament – yet we do not now teach students those skills. This skill is developed by practising producing solutions, testing them, explaining them and defending them. It calls not only for exercises in synthesising, using knowledge learnt from formal lecture courses, not only testing using laboratory methods or numerical ones, but also for beginning to develop client-handling skills, since judgement between independent variables is so dependent on broad evaluation.

On the role of the engineer:

> We must explain that we are not 'just the structural engineer' but jointly designers, bringing a knowledge of structure, materials and construction to the problem. We must claim our history – in journals, in exhibitions, on TV; we must explain the qualities we provide. We must demand the right to be allowed also to take a lead in building competitions; we must debate our educational needs and ask the University Grants Council for its support. We must examine more broadly our research needs. We must examine working abroad; we should consider how better to serve the Third World .... I nearly called this address 'Would you at the back like to come and sit in the front?' Because that is the question.

The approach that Ted Happold took as a professor at Bath was not universally welcomed; indeed he upset more than a few people. When he accepted the post it had been tacitly assumed that he would give up the bulk of his consulting work, as is usual in such appointments in Britain. But what did he do? He founded a new practice which, inside a decade, was world-renowned. At the university he was often the butt of comments about never being there; just as he was at his practice down the road. This was further aggravated when he took on his various roles at the Institution of Structural Engineers, the Construction Industry Council (which he founded during this time), the PSA and more. Perhaps the most striking example of Ted's flouting convention arose when he was faced with some complaints from students:

> In 1982 the engineering students of Bath University questioned why their course was both harder and different from the courses of their contemporaries at other universities and ... expressed their concern that, unlike their architectural student contemporaries, they neither knew who were the important designers in their discipline nor had ever heard them talk about their values and methods.
>
> I offered to close the school for three days and to try and seduce some of the greats of engineering to come and talk to them. The students and staff conceived a student conference, free to all comers. They wrote to all schools of civil and services engineering in the country inviting them. The replies were varied. Some welcomed it enthusiastically, some

asked if we could provide money for them to come! Some even said that the students would miss two days lectures and the staff could not allow that. In frustration we sent the poster [shown opposite] out – it became a cult object. Over three hundred students arrived, most bringing a favourite lecturer with them. What evolved from the initiative was a conference for students on the theme 'The Nature of Engineering Design'. Lecturing went on for ten hours continuously each day to a packed and enthralled audience. Discussions and design exercises, critted publicly, went on simultaneously.

The list of speakers contained most of the famous names of British civil engineering. Such as Oleg Kerensky, Paul Back and David Lee gave the students an opportunity to hear the views of people who designed the Kariba dam, the Severn and Forth bridges, the Sydney Harbour Bridge and so on .... Paul Back, the dam designer, explained how he tried to work with nature. Derek Sugden discussed the balance between engineering intuition and architectural precedent. Tom Maver showed the latest computer aided design techniques. John Derrington took everyone through the decisions of a concrete gravity platform for North Sea Oil. James Gordon ... argued for formal education but including a wider focus, seeing the whole rather that parts. Also Fritz Leonhardt and Fritz Wenzel came over to take part.

One thing was clear: very few engineers are asked to philosophise on their work. As Stefan Tietz said, he felt like a centipede trying to describe how he walked – 'I know how I got here but I am not too sure which leg I moved first.' Yet the attempts fascinated the students.

It is significant that Ted presented this story at a conference of the International Association of Bridge and Structural Engineers. For more than twenty years he was a regular attender at the conferences of IABSE and of the International Association for Shell and Spatial Structures. Curiously, or perhaps not so curiously, neither of these organisations is well patronised by British engineers, in comparison to other nationals; but for Ted, they formed an essential part of his worldwide social network of engineers. For more than a decade he was a regular participator (including a spell as chairman) of IABSE's Working Commission 5 Organisation of the Design Process and he chaired the organising committee for the Group's 1981 conference in London, The Selection of Structural Form. These conferences were an ideal forum for the more philosophical subjects that he so enjoyed discussing, although the Germans, in particular, could never quite make him out 'a bit of an eccentric with some funny views', as one said, with affection. He was more of a showman than many of them could handle and particularly enjoyed confounding people's expectations of engineers as rather serious, sober people – he arrived with two passengers in his Porsche at one conference in Zurich. Ted always enjoyed bringing up subjects that seemed to him, though not always to his audience, of vital interest and importance to engineers. Invited to give a paper at an IABSE conference on Long Span Structures, he talked about mosques from the world of Islam. It was a great tribute to Ted's activity in both of these organisations that the IASS conference on The Conceptual Design of Structures at Stuttgart University in October 1996 was dedicated to the memory of 'our colleague and friend, the great structural engineer Sir Edmund (Ted) Happold'. His paper on Conceptual Design, read by Ian Liddell at the start of the conference, was the last thing he had written, a few weeks before he died.

WHO'S WHO ON THE POSTER :

- Robert Stephenson: 1803–1859. Son of George with whom co-builder of locomotive The Rocket. London-Birmingham-Holyhead railway, Britannia Tubular Bridge, High Level bridge, Newcastle.
- Abraham Darby (III): 1750–1791. Ironmaster. Fabricator of early steam engines. Client and fabricator of the cast-iron bridge at Ironbridge, 1779.
- Batty Langley: 1696–1751. Author of the classic The Builder's Jewel which ran to more than a dozen editions. Vociferous critic of design and construction of Labelye's Westminster Bridge.
- William Etheridge: c.1705–1776. Piling for Westminster bridge. Devised improved method of sawing off timber piles under water.
- Robert Maillart: 1872–1940. Swiss reinforced concrete bridge designer and constructor
- Thomas Brassey: 1805–1870. The first great railway contractor. Built 23,740 miles in Britain, France, Italy, Spain, Switzerland, Holland, Norway, Sweden, Denmark, Canada, Australia, South America, Turkey, Austria.
- William Fairbairn: 1789–1874. Ironmaster. Machinery, ship and engine designer and manufacturer. Builder of mills e.g. at Saltaire. Worked on design and development of Britannia Tubular Bridge at Menai.
- Edwin Chadwick: 1800–1890. Social reformer, improved working conditions on sites. His report on public health (1842) led to great expansion of public works engineering in UK.
- William Jessop: 1745–1814. Canal builder, bridge, drainage and harbour engineer.
- Isambard Kingdom Brunel: 1806–1859. Thames Tunnel, the world's first tunnel under water, Clifton Suspension Bridge, Great Western Railway, Bridges at Chepstow and Saltash. Ship designer including Great Eastern and Great Western.
- George Stephenson: 1781–1846. Stockton-Darlington Railway, Liverpool-Manchester Railway. Co-builder of locomotive The Rocket with son Robert. Founded Institution of Mechanical Engineers.
- John Smeaton: 1724–1792. Early consulting civil and mechanical engineer. Designer of wind and water mills, steam engines, Eddystone lighthouse, canals, etc.
- Joseph Locke: 1805–1860. Railway engineer renowned for low and accurate cost estimates. Woodhead tunnel, nr. Sheffield, the world's longest in 1839. With Brunel and Robert Stephenson known as 'the great triumvirate'.
- Thomas Telford: 1757–1834. Founder of Institution of Civil Engineers. Prolific road, canal, aqueduct and bridge builder. Engineer of exquisite cast iron arch bridges. Pont Cysyllte Aqueduct. Menai and Conway suspension bridges.
- George Deacon: 1843–1909. Water Engineer. Vrnwy dam.
- James Jardine: 1776–1858. Water engineer. Union Canal. Edinburgh water supply.
- J E Etherington: 1806–1862. Railway engineer. Became Joseph Locke's partner.
- F Pritchard: 1723–1777. Author of first scheme design for Ironbridge, 1775.

Opposite, from top: One of the few berobed shots of Ted Happold in captivity; Happold's joint paper on The Design of Air Supported Structures, 1984; Poster for the winter conference in Bath in 1983, taking a 'time machine approach to participants'.

Work at Buro Happold

Hong Kong Ocean Park 1984

Buro Happold's work in Hong Kong was largely concerned with the export of specific skills demonstrated at the Tsin Sha Tsui Cultural Centre and at Ocean Park. Ho Happold were the local facilitators for much of the work. The work with Derek Walker Associates tended to be generated by the Royal Hong Kong Jockey Club who allocated generous funds each year for community projects. They were initially responsible for developing Ocean Park, which was spun off to private operators Ocean Park Hong Kong who commissioned the aviary. Buro Happold's expertise in large free-form aviaries demonstrated in Munich made them natural collaborators for Leigh and Orange, the architects appointed for the project. The 2,700-metre-square aviary at Ocean Park uses a 12 gauge (2.7mm diameter) mesh. The contractor elected to buy plain wire from Japan and to crimp and weave the wire and weld it into its finished fabric form on site. Because of the aviary's location close to the sea, low carbon 316 austenitic stainless steel mesh (which has additional molybdenum content) was chosen to enhance the resistance to corrosion from chlorides.

The master plan for the University of Science and Technology was a joint proposal by Derek Walker Associates and Buro Happold. It would be hard to conceive of a more dramatic site: a spectacular cascade of terraces falling 150m from south to north and then east to the shoreline of Junk Bay and the South China Sea. Geotechnical problems limited economical construction to the higher western edge of the site and this suggested that the main academic areas of the university should be focused in a relatively compact form where pre-engineered terraces were available and that the shoreline and outlying areas should be utilised for residential and sports facilities.

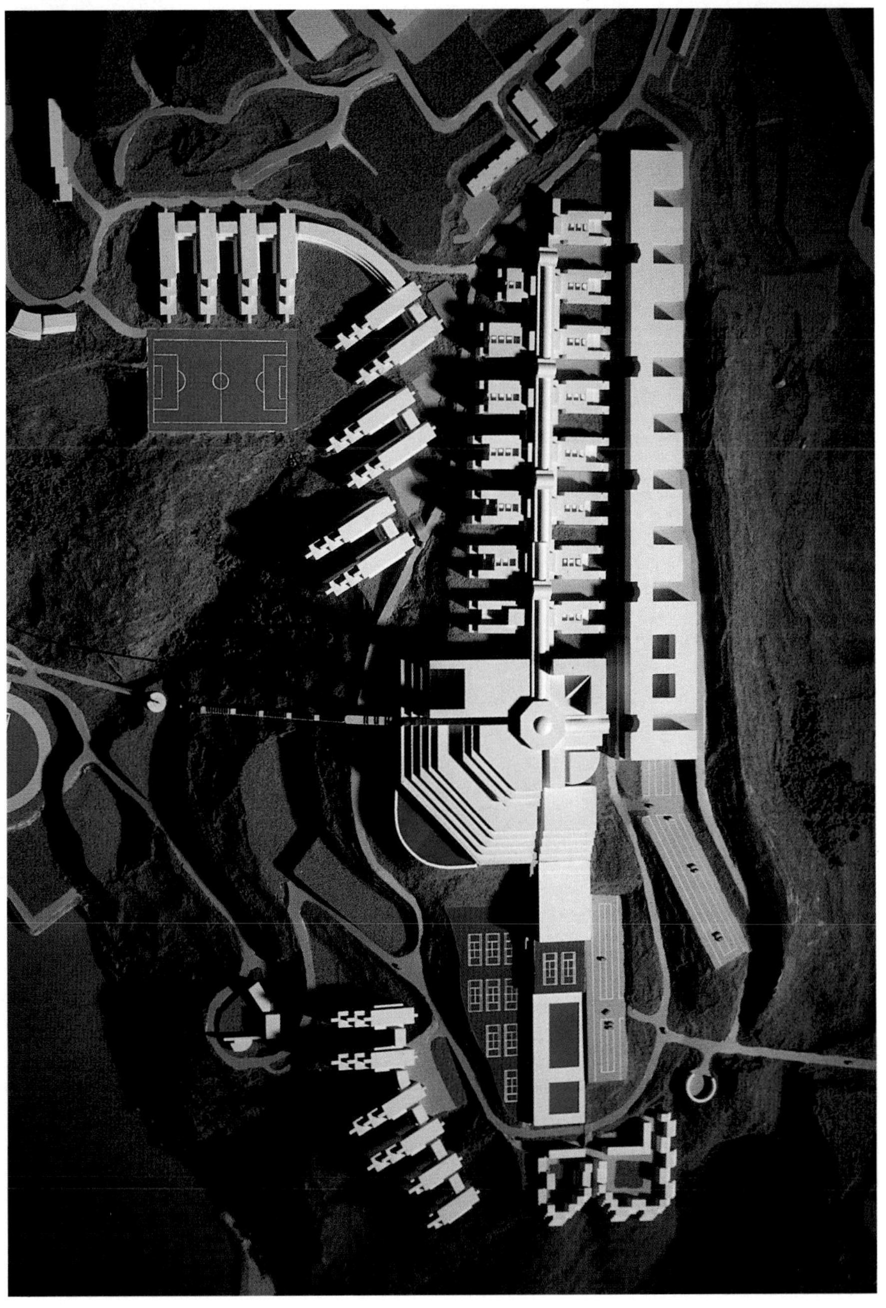

Opposite: Ocean Park Aviary; Top right: General view of Ocean Park; Right: The Chinese Technical University of Hong Kong site model.

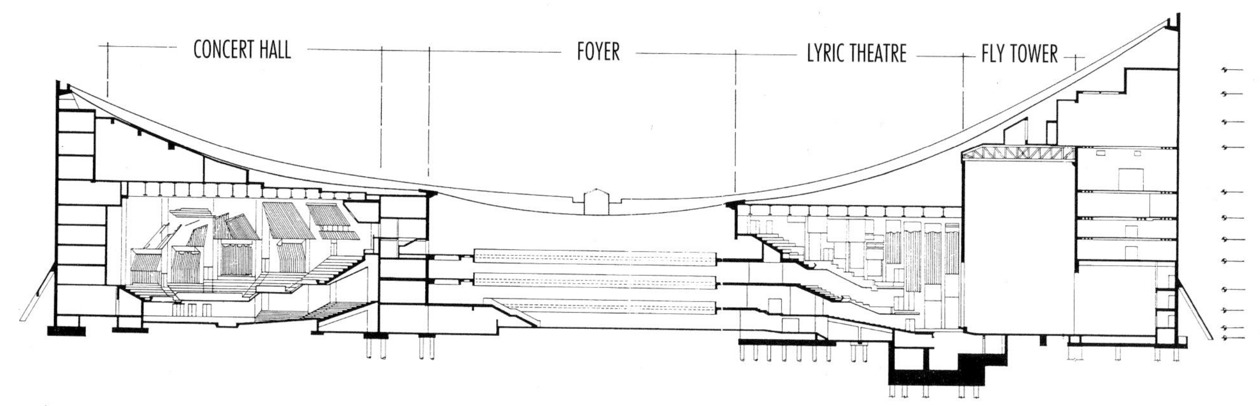

Above: Cross section through Tsim Sha Tsui Cultural Centre; Below: Construction shots; Opposite: Views of the main auditorium and foyer plus detail construction shots.

Tsim Sha Tsui Cultural Centre, Hong Kong

Completed in 1988 and officially opened by the Prince of Wales in 1989, the cultural centre lies at the southern tip of the Kowloon peninsular within Hong Kong Harbour. The architects for the complex were the Architectural Services Department under the direction of Jose Lei. The administration building, restaurant block, underground car park and plant rooms (Phase 1) were designed by the Engineering Services Department. However, when the project was reactivated in 1981 and the engineering services put out to international competition, Ho Happold were declared the winners (largely due to Ted's expressed confidence that the hanging 'ski-jump' roof form could be resolved economically by utilising a hanging grid form) and acted as civil and structural engineering consultant for the complex. The Ho Happold team included Ted Happold, Tom Ho, John Morrison, Terry Ealey, Eddie Pugh, Padraic Kelly and, as resident engineer, Mike Cook. Occupying a floor area of 30,000 square metres the complex comprises a 2,100-seat concert hall and 1,750-seat lyric theatre, separated by a multi-level central foyer with linkage to a 450 seat studio theatre. The building structure has a number of unusual and novel features, including the first barrette foundations to be used in Hong Kong. The structure has no columns and large open spaces are achieved by the use of deep beam walls, long span floors and roof construction and large cantilevers. The roof was formed from a series of three hung concrete rigidised lattice cable nets so as to be able to resist the extremely high wind suction forces associated with the typhoon winds of the South China Sea. The tension forces from the roof are resisted by substantially reinforced concrete ring beams and the building below relies on the liberal use of deep wall beams spanning over relatively large volumes.

SWIMMING POOL
QUEUE-UP AREA

KOWLOON PARK
OPENING
CEREMONY

Kowloon Park

Derek Walker Associates were the architects, landscape architects and planners and Buro Happold the civil and structural engineers for this 30 acre park in the crowded Tsim Sha Tsui area of Kowloon. In Hong Kong this kind of facility is in short supply, and the park now welcomes about 100,000 visitors a day. The main task was to accommodate within a very limited space the needs of a very broad spectrum of users ranging from the young, active and noisy to the old and infirm. Proposals were based on an arrangement of outdoor rooms supported by a network of routes using a strong north–south axis leading to the sports centre. Attractions in the park include an outdoor arena and performance area, an aviary, a Chinese garden, a children's adventure playground, a sculpture park, water gardens, a bird lake, scented gardens, a continuous loggia, a history museum and maze. The sports complex at the north end of the park has a high barrel-vaulted central entrance concourse with a cross-axis leading to the Olympic swimming complex to the east and sports halls to the west. The 50m pool has seating for 1,500 and there is also a recreation pool, practice pool and diving pool. Other features include a Banyan Court, soccer pitches and a series of free-form pools and cascades. The main building features a coffered reinforced concrete barrel vault roof on pre cast tie beams over the main entrance concourse. The wide spans of the swimming pool and sports hall are a family of site-assembled steel lattice constructions. The longest truss weighs 10 tons. After assembly the trusses were painted, lifted in one piece and slid along the eaves beam to their final positions. The building, clad in white enamelled steel panels, has developed a language of glazing in white metal frames.

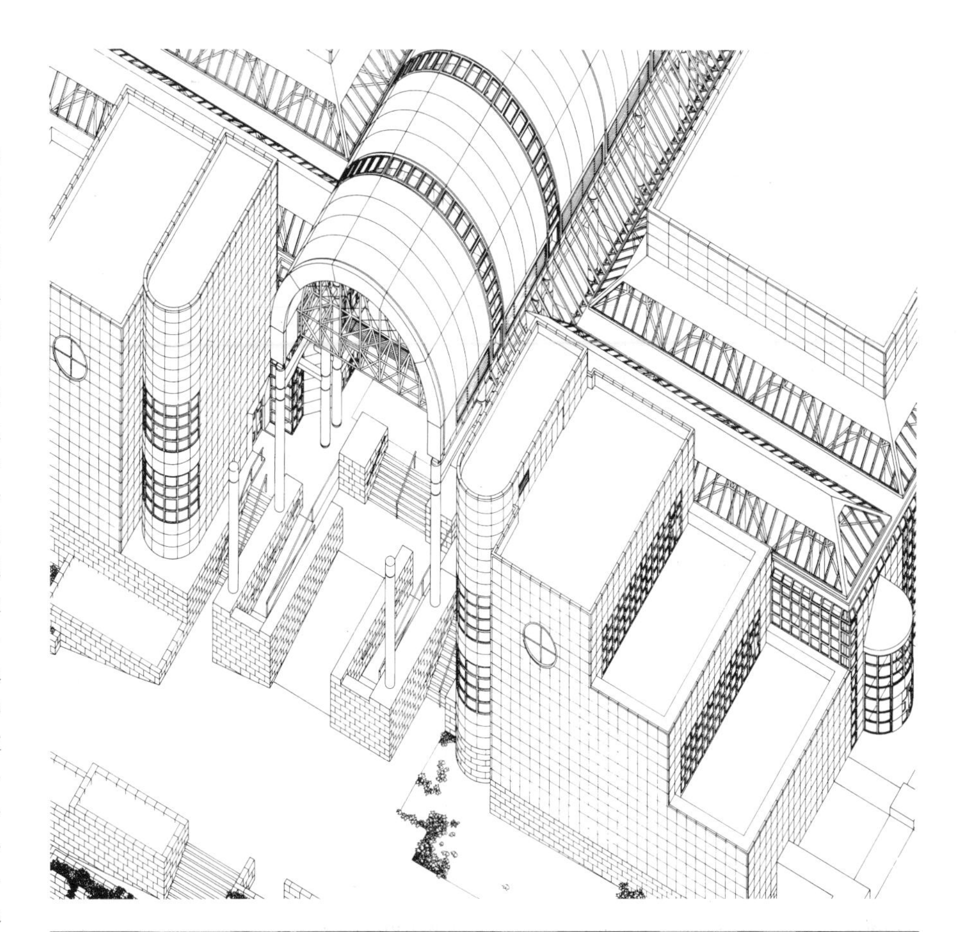

Opposite, clockwise from top left: Kowloon Park, Banyan Court; Bird Lake, Structure Entrance Concourse; View of sports building by night; Interior of fun pool; Olympic Pool; Sculpture Walk; Detail of pedestrian bridge.

Above: Axonometric of the concourse area and site model of the park.

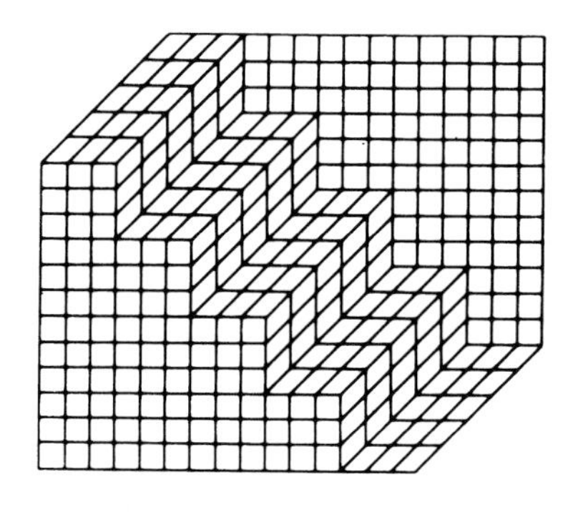

Theo Crosby Pentagram

Theo Crosby, one of the founding partners of Pentagram, was another lifelong friend of Happold. Crosby saw in Ted the epitome of the restless and creative scientist, a man of similar wide interests who could not be pigeonholed into any particular engineering category. To Theo, Happold was the ultimate creative catalyst whose ability for lateral thinking could also turn an argument or redirect a design process. Theo, as advisor to the Union Internationale des Architects and the Saudi Government, suggested Dannatt and Happold should enter the Riyadh competition for the conference centre. He turned to Ted for co-operation in the design of the British Genius Tent at Battersea in 1977. His partners, Kenneth Grange and Alan Fletcher who (designed the Happold corporate logo) arranged a Happold tour de force for the Aspen Design Conference in 1986. Kenneth Grange was also responsible for nominating Ted to the Faculty of Royal Designers for Industry and he also involved the Buro in a suite of street furniture to repulse an imminent French invasion. It was also significant that the long-running crusade by Sam Wannamaker and Theo Crosby to recreate the Globe Theatre was engineered by Buro Happold, stimulated by Ted's appetite for historical research, which was as insatiable as his obsession with the future of engineering.

Left, from top: British Genius tent interior detail; Exterior view; Pentogram bus shelter; Above: Buro Happold logo; Opposite: Two views of the Globe Theatre, London, in construction.

Michael Hopkins

Following the abortive project to create a gigantic lightweight umbrella to enclose the immediate environment of the 200m x 36m town square in Basildon, Buro Happold worked with Michael Hopkins on the Fleet Velmead School in Hampshire. Following the successful completion of this collaboration, the practices consolidated their relationship on the construction for Solid State Logic of a 45 x 45m building housing research and development laboratories in an elegant two-storey pavilion. The moulded floor slab follows a 7.2m x 7.2m grid and sits on simple metal columns. Servicing is enclosed in the raised floors. The first floor has a space frame roof on a 14.4m grid which overhangs the ground floor for shading. The architectural theme integrates structure and services, and in due course this elegant, self-standing, glazed and shaded pavilion led to the somewhat more elaborate research pavilions and central atrium for the extension to Schlumberger. Here the added thermal mass is achieved by the permanent ferro-cement forms, the roof structure is anchored by spheroidal steel castings and the atrium is clad with ETFE cushions first considered by Buro Happold for 58º N.

Ground floor ceiling ribs.

❍ Ground floor columns, ●First floor columns.

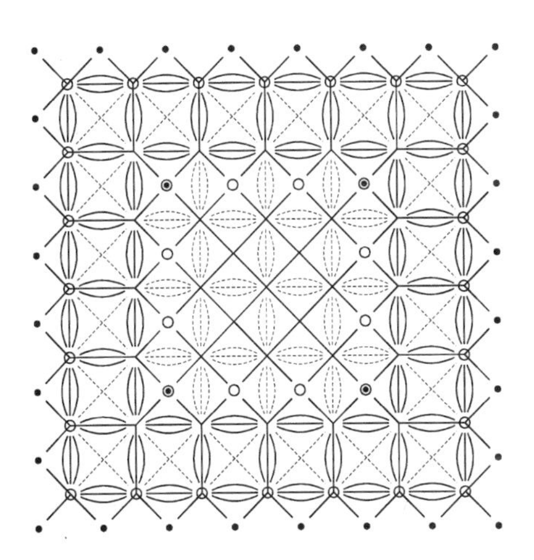

Primary structure ribs., Secondary structure ribs., Structure of first floor slab

In 1988 the joint venture of IBM/MEPC commissioned Michael Hopkins and Partners, Edward Cullinan and Buro Happold to develop an 18-acre commercial site with some 400,000 square feet of offices with 1,640 car parking spaces in a landscape parkland at Bedfont Lakes. The commercial site was retrieved from a 250-acre waste tip being redeveloped as parkland to the south of Heathrow. The master plan envisaged a landscaped square around which a number of fine modern buildings were located, with two levels of underground car parking beneath central landscapes. In the south overlooking a new lake is a modern steel-framed building with externally expressed steel façade and cast steel nodes. The office structure itself was carefully designed as a fire–protected exposed steel structure, using a composite steel and precast plank floor. This thin floor structure accommodates the clear service void within a minimum depth for the ATM air conditioning units enabling individual control in each office. Internally located is a 54m x 18m clear span doubly glazed atrium, suspended from tubular lattice steel structures and shaded from excessive solar glare by external PVC polyester hyper sails. Beneath the central square are two levels of car parking which are naturally lit and ventilated to create a user-friendly and relatively clear span environment constructed of purpose designed, precast concrete frames and carefully detailed retaining walls.

Most recently the Michael Hopkins/Buro Happold team completed the Queen's Building in the grounds of Emmanuel College, Cambridge. This little building for a 200-seat concert hall designed to the highest acoustic standard is contained within a carefully crafted solid stone structure using load-bearing Ketton stone, the same material used for Sir Christopher Wren's adjacent seventeenth-century chapel. Solid stone columns are post tensioned with a single 32mm stainless steel rod anchored by way of cast-in stainless steel nodes. The vertical stone columns, 800m x 400mm at the base tapering to 600m x 400mm at the third storey, are post tensioned with 25 tonnes force to achieve continuity of structure and to prevent tension cracks developing under eccentric loadings. Internally the quiet acoustic space of the auditorium, enabled by Derek Sugden of Arup Associates, is roofed with a heavyweight timber and cement pugging supported on trusses of stainless steel and American white oak supported off corbels on top of the stone frames. Natural light into the auditorium is provided by triple glazed windows at the clerestory level above non-load-bearing ashlar panels infilling the stone frame.

Opposite: The designs of the Schlumberger building. Michael Hopkins and Buro Happold were partly inspired by Pier Luigi Nervi, the engineer who in the late forties first exploited the idea of permanent form work and developed a suitable material from which to make the forms – ferro-cement is a composite made from thin mortar with steel mesh reinforcement. It tends to be more costly but continuously varying profiles can be achieved more easily with pre-cast permanent form work than with conventional in situ concrete. To exploit this to the full, a complex though regular pattern had to be conceived to resolve several conflicting influences. The rib pattern also needed to take account of three different grids in the floor plan – a column grid at ground level, the column grid at first-floor level, and the grid for lighting and air conditioning.

Opposite, top three images: The Queen's Building, Emmanuel College, Cambridge. 200-seat concert hall, solid stone structure using Ketton stone. The structural solution was to stabilise the stone frame by placing 32mm-diameter stainless steel rods through each of the buildings 28 stone columns and post tensioning them with a vertical 250kN force. Internally however the heavy lead cover roof is supported by a composite truss of white American oak rafters and meticulously machined stainless steel tie rods, plates and Queen post.

Bottom three images: Solid state logic. New facilities at Begbroke, Oxford. A high-tech company with the building placed in a high-quality landscaped setting. 3,500 square metre of usable space is provided in a two-storey building. At its centre a naturally lit atrium (11.6 square metres linked to the ground floor (40.8 square metres) by means of a diagonally placed metal and glass stairway. The first floor overhangs the ground floor by 1.8m. A fair faced off-white in situ concrete slab 375m thick provides a permanent sandblasted gridded soffit on a 1.8m module. The heavily insulated flat roof for the first floor and atrium is achieved by elegant modifications to the standard space frame.

This page: New Square Bedfont Lakes, Surrey, for the IBM headquarters. The architects wished to create a building with an exposed steel frame. They were also keen to retain the metallic crisp surface texture. Throughout the building the columns reflect the size and load they carry, and the transition between different sections at different floors is pointed up by an ornate steel casting.

Worcester College

From first appearance this building would seem to be of conventional domestic construction. In fact the design of both its internal environment and its structure is much more complex and each is intimately integrated with the other. The building form and structure arose out of a wish by the architects, McCormac, Jamieson & Pritchard, to develop a highly energy efficient building. This led to a compact floor plate and the grouping of a number of student study bedrooms around a shared kitchen, living room and bathroom. This group of rooms is repeated and stepped back in each of the building's three storeys to create terraces overlooking the lake and college playing fields. The layout of spaces within the building led to an environmental hierarchy, with the warmest spaces insulated from the outside by the two cooler spaces in order to minimise temperature differences across walls. Semi-public corridors are maintained at low temperature in winter and private internal corridors – linking study bedrooms, kitchen, living rooms and bathrooms – are maintained at normal room temperature of about 18°C. Student rooms, a large proportion of the building area, are maintained at 13°C which can be increased when occupied. The energy needed to achieve this environmental performance was minimised by using carefully chosen structural materials. The walls of the internal constant-temperature spaces are built of concrete and blockwork with a high thermal inertia and heat capacity. The walls between student bedrooms are of dense concrete blockwork to give good acoustic insulation, while their external walls are of brick and light cavity wall construction. Both walls and ceilings are lined with a relatively lightweight plasterboard backed with insulation to allow rapid heat gain during occupancy. These environmental influences had a direct bearing on the design of the load bearing structure which comprises reinforced concrete floors supported by masonry walls and piers, a type of construction in which dead loads dominate the relatively small live loads.

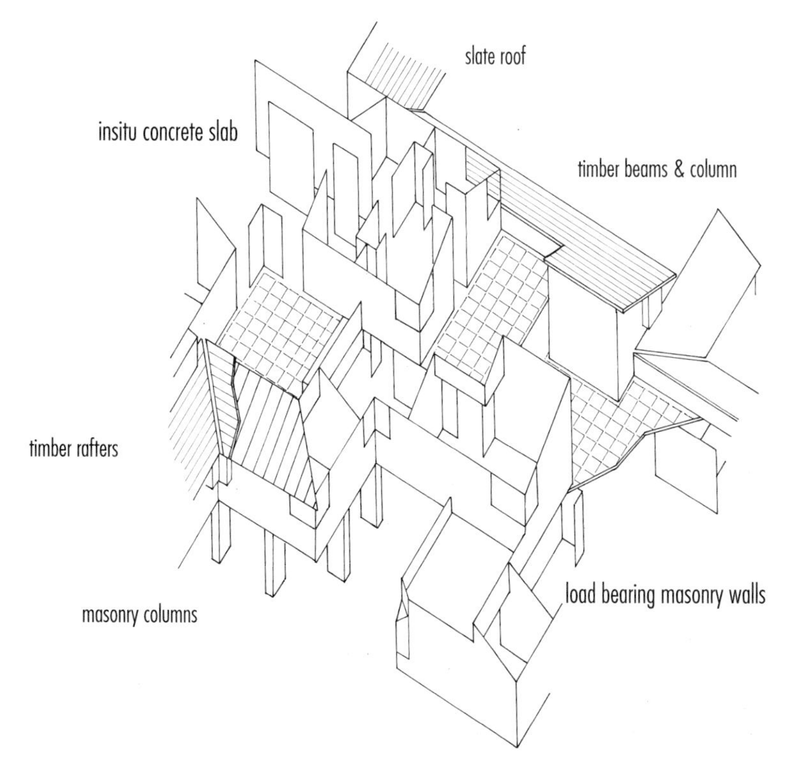

Opposite, top: Terracing and stepped levels of the Sainsbury's Building; Bottom: View across the lake.

Above, from top: Common room with view onto the terraced garden; Axonometric showing building levels and main structural features and the complex load path from roof down to ground level.

Hampshire Schools

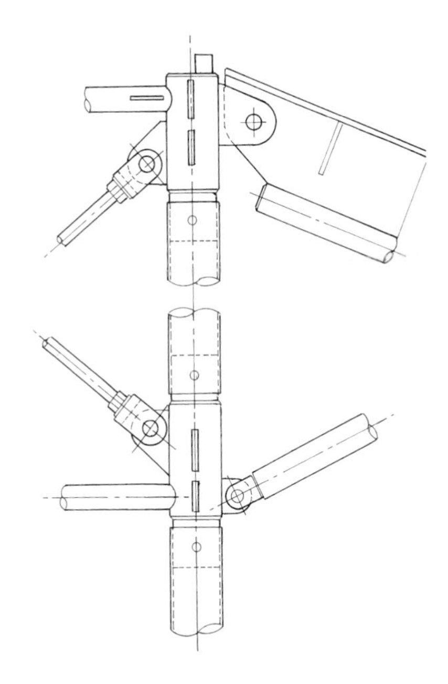

The work of Sir Colin Stansfield-Smith's group at Hampshire County Council during the eighties provided the most interesting new thinking in school design since the fifties. A comparative study of three schools – Fleet Velmead (1986) by Michael Hopkins and Partners, Bishopstoke (1988) and West Totton (1989) by Hampshire County Council show graphically how their form finding was driven by interesting engineering solutions.

Being infant schools operating mainly by day, each school's general arrangement is generated by the need, through its section to introduce natural light deep into the classroom plan using central circulation as shared storage and resource areas. All three schools utilise this possibility in interesting but different ways in the organisation of both plan and section. West Totton is on a small flat site surrounded by housing. The side blocks of the gridded H plan contain the classrooms which look inward. The exposed planed timber grid roof construction exploits the economical possibility in modern laminated timber by using small pieces of wood, accurately prefabricated in order to build up a stronger, stiffer composite structure – one which in its completed state is easily understood by the user. At the same time, the roof structure provides a visually pleasing but organised construction system from renewable materials. Mechanically laminated timber structures date back to the early ninettenth century and the railway arches designed by Brunel. However, it was not until the 1940s. following the development of powerful synthetic resin adhesive that glued laminated timber (Glulam) began to emerge. Glulam was used consistently throughout the building not only for the structural frame but for elements including straight and cranked window mullions, floor beams and staircase. The initial design required the assembly of 10m x 15m roof sections on the ground slab, each roof section being given a permanent pre-camber.

Bishopstoke is very different, the building being planned around the central hall, a toplit drum 15m in diameter which generates the radial spiral grid of nine classrooms and also accommodates the administration areas. The classrooms look outward beneath the eaves to the playground protected by earth bunds. The most striking structural feature is the timber roof, an effect largely achieved by the buildings' geometry. From the glazed apex the roof line curves down over the hall until it meets the spiral grids of the classrooms below. From this the gradients become progressively shallower. The geometrical effect of continuing these two grid systems produces some very complex shapes and places for although each ring of purlins is set to a single level the length of each rafter between purlins progressively increases, resulting in a surface that is in fact part of a helix.

Fleet Velmead takes a different approach. The open plan school generates a low key but highly engineered appearance as a pavilion structure utilising high quality steelwork details. Despite being constructed of unashamedly modern materials, the building and its playground try to make the minimum impact on the heathland site. Indeed, the full size window screens seek to bring the site into the classroom. In concept the structure has its origins in those large French electricity pylons that support their wires on outriggers. Here the tubular steel corridor frames (76mm in diameter with 5mm racking steel tubular struts) support the roof rafters at 6m centres thus reducing the 10m span of the inclined roof rafters to a continuous two-span beam of 2m and 8m respectively.

Opposite: View of the exterior and interior of Fleet Velmead Infant School. Stressed PVC sails form solar shades over south-facing classroom perimeter. The interior is dominated by a top-glazed central corridor allowing light deep into the school. Lateral stability is ensured by the stiffness of the braced H corridor forms in conjunction with other finishes. Detailing of the structural steel work and panel system was considered important enough to fully articulate to the user each structural part.

Above: Details of the main structural junctions , Fleet Velmead.

Page 120: Bishopstoke Infants School showing erection of the rafters, finished apex from within the school, the dramatic effect of the finished school building and the interior and section showing the quality of the structural system.

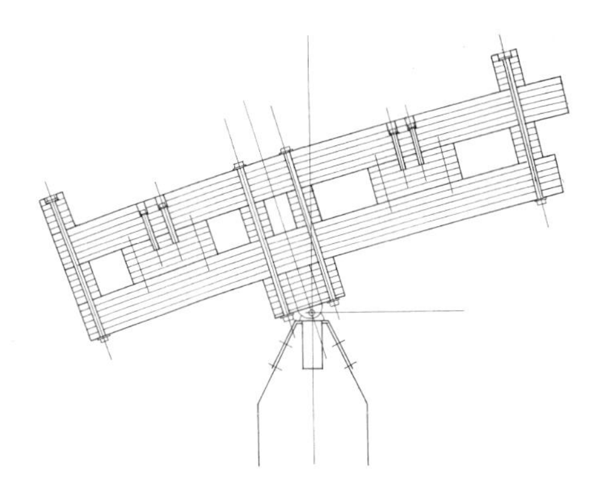

Detail of stainless steel pin joint between roof and column.

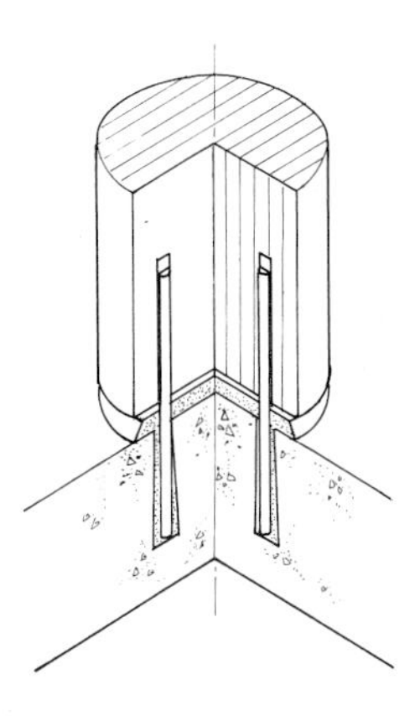

Rigid connection between Glulam column and floor slab.

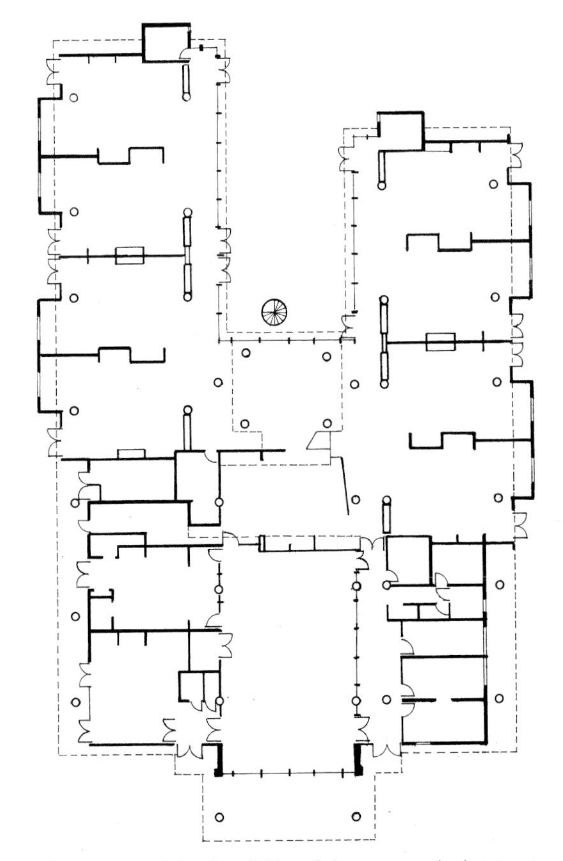

West Totton. Above: Structural details and Plan of classrooms and administration; Right, from top: Manoeuvring roof sections on to columns; interior of school hall; exterior view; Glulam timber construction.

Diplomatic Club, Riyadh

Tuwaig Palace is undoubtedly the most significant and expressive building in the Arriyadh Diplomatic Quarter. The Palace was conceived as a primary meeting, entertainment and social centre for the diverse interests of the very cosmopolitan diplomatic community. Located on a promontory overlooking the picturesque Wadi Hanifa, the external elevations of the Palace have been designed to harmonise with the rocky plateau landscape. The curvilinear plan of the Arriyadh limestone walls followed the perimeter of the 75,000 square metre site. The built form surrounds a soft oasis containing lawns, exotic plantings and over 100 fine palm trees. The massive stone-clad curving double concrete wall enclosing the accommodation was designed to receive the anchorage forces from the external PFTE glass membrane tension desert rose roof and the two internal cable nets which are insulated and tiled. The three floors within the curving double wall are structured with precast concrete stub girders acting compositely with the floor slabs – reminiscent of the composite palm and roundwood timber roofs of the region. As the garden centrepiece the consultant team created the 17m-diameter Heart Tent, a regular conical 326mm-mesh cable net suspended from a 7.3m-high central lattice mast and ten 2.3m-perimeter stayed tubular masts. All structural components are of stainless steel; the central mast consisting of five curved stainless bars and welded lattice bracing. The outer mast tripods are tubular compression members supported on exposed aggregate precast concrete pedestals. The canopy itself is made up of over 2,000 8mm-thick tiles of float glass. The geometric shape requires that the glass tiles are square or rhombic in shape. The tile artwork was created by Bettina Otto, based on images of foliage and blossoms. The background colours of the ten segments are in graduated warm spectrum tones with the densest colours used in the southern aspect.

Opposite, top: Heart Tent; Bottom: General view of the Diplomatic Club.

Above, clockwise from top: The massive stone-clad double walls within the Club; Heart Tent glass design painted by Bettina Otto, 1988; View from below the semicircular window that forms the upper conclusion of the membrane and the covering elements for connection with the wall; External View of the Club. Junction detail of the stained glass with the column Heart tent.

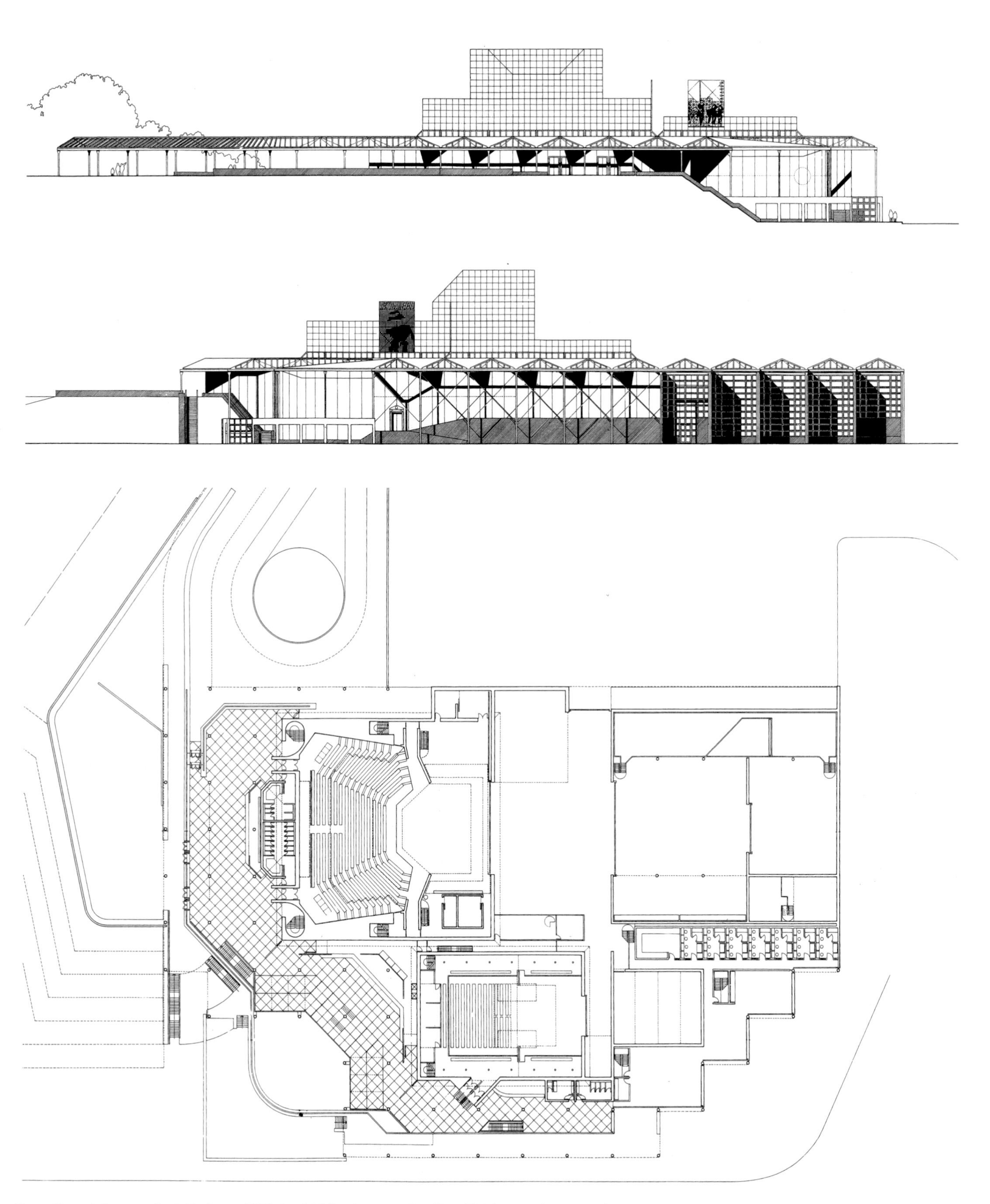

Above: Plan and elevations, Leeds Playhouse, 1985. Derek Walker Associates, Buro Happold, John Bury Theatre consultant.

Theatres

In ancient Greece the space adjacent to the market place at 'Agora' became known as the 'Orchestra' and gradually evolved into a focus between actor, chorus and audience. Gradually earlier timber forms were superseded by steeply tiered stone fan-shaped seating giving shorter site lines and better acoustics. Then followed the 'skene' to frame the performance space, and thereafter the proscenium. In Europe two built forms have been evolving that are specific to the theatre environment. Diverse forms such as courtyard, proscenium, picture stage, full thrust, semi thrust and in the round have all been part of the aim to improve communication between performer and audience. That was why the reconstruction of the Globe with Theo Crosby was so interesting in placing a single period in context. Today, of course, it is vitally important that this build form can be operated, managed and maintained by the minimum of personnel so reduce running costs. In this context Happold's interest in theatre was stimulated by his work in the seventies with Renton Howard Wood and Levin on the Crucible at Sheffield, and with the same architects (using John Bury as theatre consultant) on the Warwick University Arts Centre.

This interest was given a more personal direction when he became involved in the development and management of the Theatre Royal, Bath, then under the Chairmanship of Jeremy Fry. The renovation of this fine provincial theatre in the early eighties tried to meet the economics of present day management. Despite containing historically important early theatre machinery, the entire Georgian fabric of the 900-seat theatre needed refurbishment and new stage and flying facilities. The front-of-house facilities were admirably planned by Donald Armstrong and Carl Toms, but the real architectural problem was the incorporation of the new fly tower and enlarged stage area. The solution was a slightly asymmetrical fly tower shape. By cladding it in lead and restoring the delicate Georgian stonework below, building mass was minimised. Appraisal of the Georgian fabric had identified that the existing stone walls could not support the increased vertical loads. The solution here was to insert a new structural steel fly tower as a rigidly braced tube within the existing wall on new foundations. Ted eventually followed Jeremy Fry as Chairman of the Theatre Board.

The experience built up in the seventies and eighties in his persistent search for a 'perfect theatre' was augmented on work on two national competitions, with Derek Walker Associates and John Bury. Both theatres – at Leeds and High Wycombe – formed part of a design continuum, starting with Warwick University. The location of the Leeds theatre in the noisy city centre required the theatre spaces to be controlled as a box within a box. The walls provide sound reduction of at least 45db and the concrete roof system, with a minimum mass of 2.5kNm≤ and supported on deep steel trusses, is designed to achieve the 20–25db noise criteria in the theatre with a reverberation time of 1–1.2 seconds. At High Wycombe the same principles of theatre design and construction were utilised to provide direct access from the town centre into the facility without acoustic interference. The most interesting development was in the 1,000-seat multi-purpose hall which was driven by an ingenious and flexible acoustic system permitting musical performance at the highest level, as well as live performances, exhibitions and banquets.

Iain McIntosh, a close friend of John Bury's and the Principal of Theatre Projects, harnessed Buro Happold's theatre experience and introduced them to Dunbar Naesmith for the Theatre Royal, Glasgow. Currently the Buro are working with the same practice on the Edinburgh Empire, the Inverness Eden Centre and the Hippodrome, Birmingham.

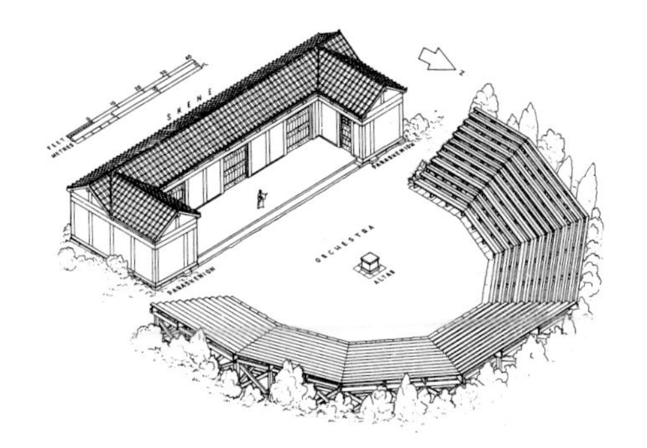

Above, from top: Early theatre in Greece from R & H Leacock, *Theatre and Playhouse,* 1984; Globe Theatre, London – Sam Wanamaker, Pentagram (Buro Happold); Warwick University Arts Centre (Renton Howard Wood Levin, Theatre Consultant, John Bury, Ove Arup & Partners).

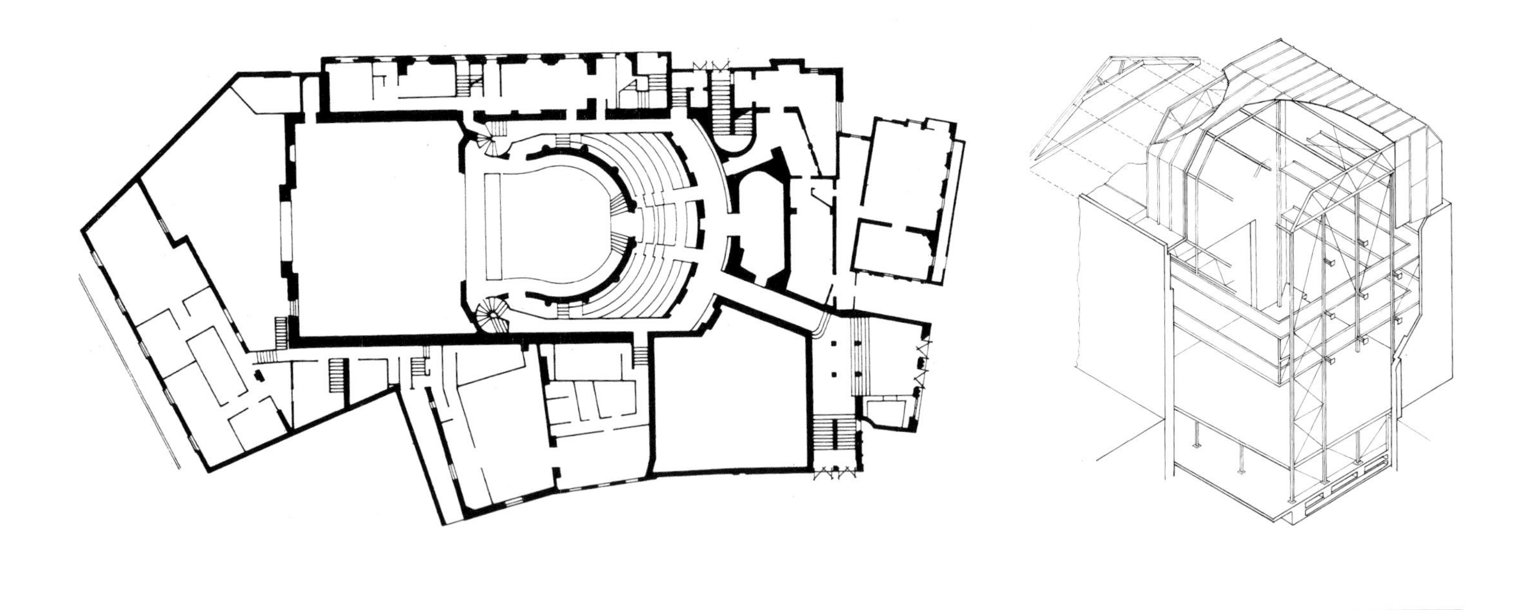

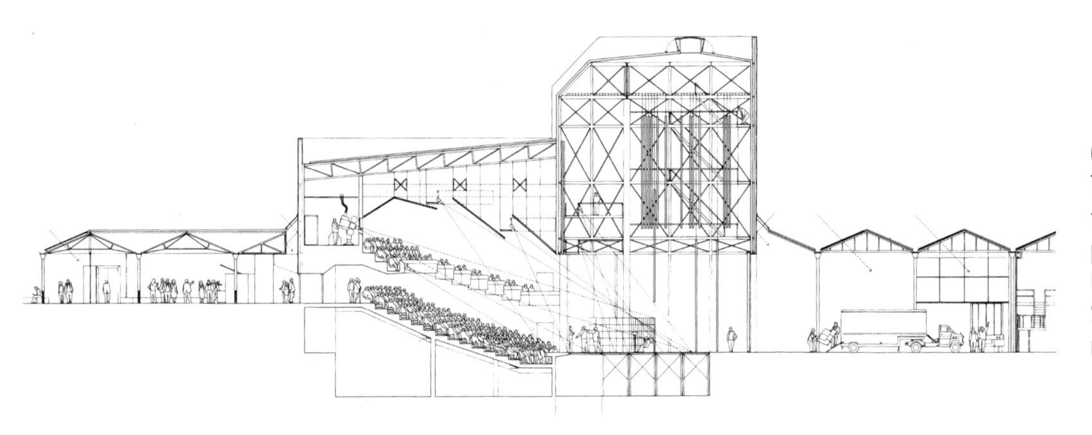

Clockwise from top left: Plan of the Theatre Royal, Bath; Isometric of new fly tower, Theatre Royal, Bath; Ground-floor plan of Warwick University Arts Centre; Plan of High Wycombe Arts Centre; Section through Leeds Playhouse showing main auditorium and stage.

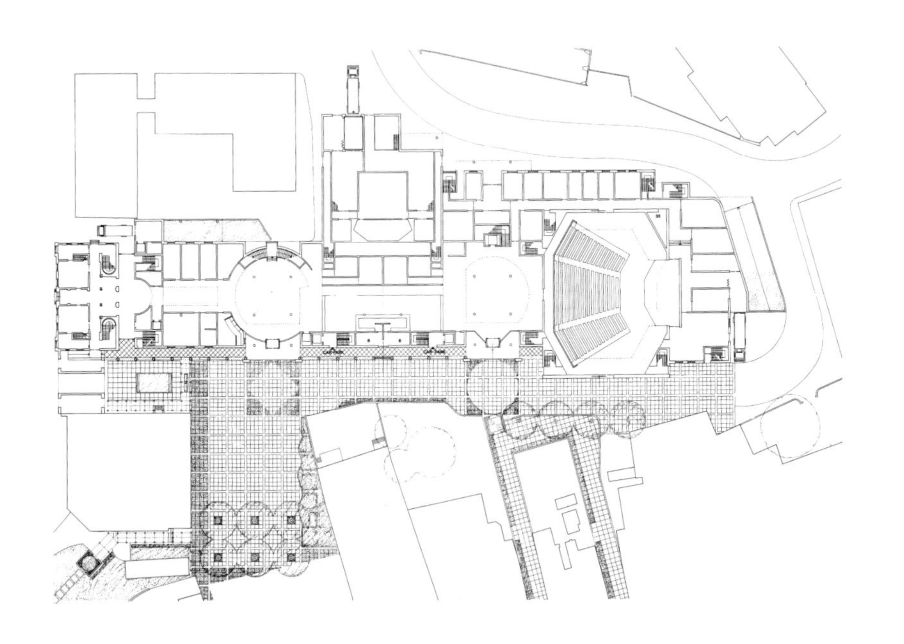

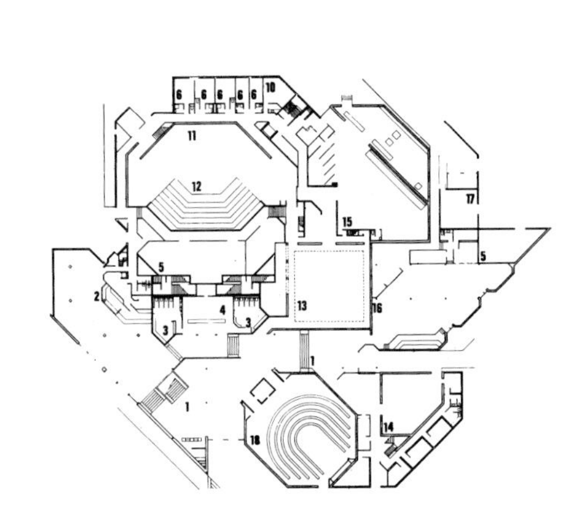

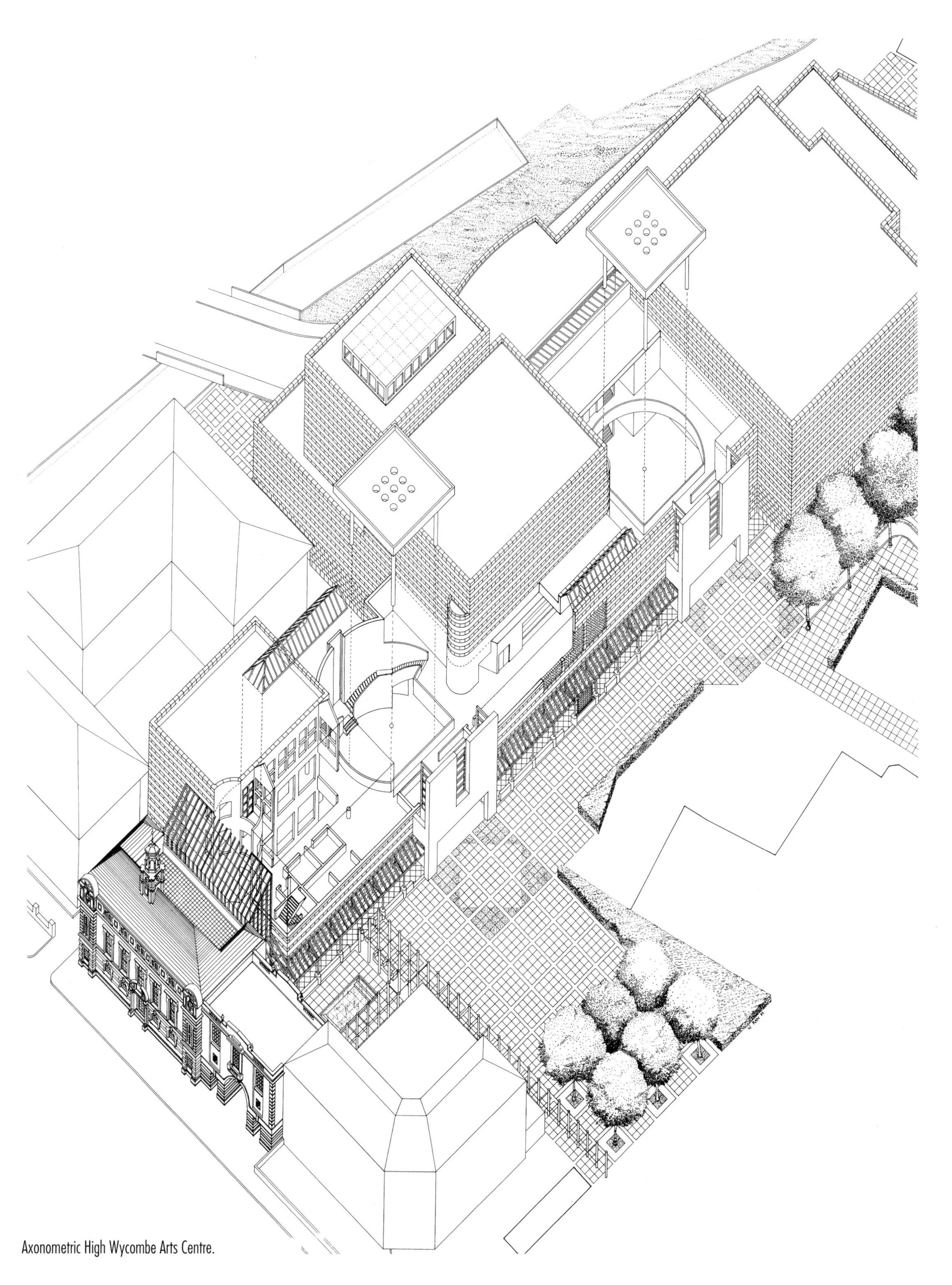

Axonometric High Wycombe Arts Centre.

ABK

Richard Burton described working with Ted as rather like witnessing a magic show – focusing on the most important element of the problem, then an instant deluge … or later a conclusive phone call or a major new idea liberated by the choice of additional material. The significance of Hooke Park and the Burton, Happold, Otto triangle is described in the next chapter. Two other ABK projects that represented important milestones were the abortive competition for British Telecom headquarters in Milton Keynes, and Techniquest in Bristol. For the former, a very large office complex of 32,800 square metres was proposed in a single storey building sandwiched into four pavilions, linked by glazed pedestrian streets with a fifth pavilion incorporating main entrance and specialist facilities. The design aimed to meet the stringent demands of information technology, creating a new form of flexible office planning and energy systems employing conservancy measures which would have resulted in a energy saving of up to 40%.

At Techniquest, within a crafted industrial envelope, a visitor attraction concentrating on interactive and scientific programmes has been created. The original ship repair shed of wrought iron and cast iron columns was restored and new foundations provided. Special structures were developed for the large south-facing glazed wall to the exhibition area and to the east and west gables. Internally, careful environmental design and integrated building services complement the hands-on science exhibition.

Another proposal from the Richard Burton, Frei Otto and Happold team was for a stained glass roof hooked to the walls of St Peter's Church, Bristol, with an intervention by Alexander Beleschenko. It is hoped that this modest project will be realised, allowing a permanent venue in Bristol for lunchtime poetry readings, talks and small orchestral concerts.

Opposite: Structure and interiors at Techniquest, Bristol.
Top: BT headquarters, Milton Keynes; Bottom left: Hooke Park; Bottom right: St Peter's, Bristol.

Design Research and Development

Research and development in the construction industry has always been a poor relation to that activity in the mass production and process industries such as electronics, aerospace and chemicals. The reasons for this are complex but the fact remains that the various players in the game either have little to gain from research for their own firm, or do not have the turnover to invest in it. Most research in construction has to be squeezed out of tight budgets – a lesson quickly learnt by Ted Happold during his days running Structures 3 at Arups. There he had helped to build up the firm's reputation for tackling unusual structures, many of which were achieved only because research, innovation and development were incorporated as an integral part of the project. During the eight years he ran Structures 3, Ted and others saw the remarkable results that could be achieved by this means – the many tension structures with Rolf Gutbrod and Frei Otto, Beaubourg, the timber lattice dome at Mannheim, and others. Some were achieved largely because the team of engineers in Structures 3 was confident they could happen. Such confidence is surprisingly rare in the construction industry where engineers are usually, extremely cautious and conservative in what they propose, especially given the short times and low budgets allowed for projects.

During the 1970s and 1980s the gap between universities and the construction industry widened. Much university research in construction engineering became too remote from the needs of design engineers. It was a vicious circle – the greater the gap, the less inclined were practising engineers to try to work with universities. And the gap was great. Even now, structural engineers with PhDs often omit this qualification from their business cards and CVs submitted to clients when trying to win contracts. They do not want to be judged 'too academic'.

During the 1990s this situation has started to turn around. Much research and development in construction engineering, both government- and privately-funded, is now being linked more closely to the needs of the industry. These changes represent a significant shift in the whole culture of research and development in construction. They came about gradually over many years and by many means and are the result of the constant campaigning by a great many engineers. Prominent among them was Ted Happold in his various roles with the Construction Industry Council, the PSA, as President of the Institution of Structural Engineers and, not least, his activity at the university of Bath.

When Ted arrived at Bath in 1976 one of his first goals was to set up a research project which clearly focused on the needs of design engineers. The subject he chose was air-inflated structures – the archetypal tension structure – which exploited the relatively newly available membrane materials that are so familiar today. Mike Cook, then a young graduate at Buro Happold, began with a Europe-wide survey of existing air-houses and their manufacturers, and the manufacturers and testers of the new membrane materials. This groundwork enabled Ted to get a large research grant from the Wolfson Foundation and to build the multi-disciplinary team he wanted to tackle the subject.

Opposite, lower images: Munich Aviary (Jorg Gribl with Frei Otto, Buro Happold); The Jorg Gribl/Frei Otto 'Structure Like a Cloud' to cover 4500 msq; A 3.2mm crimped stainless steel wire form the mesh; Aviary under construction; View of the Munich Aviary; A mast head connection, Munich; Upper images Hook Park Workshop (ABK with Frei Otto, Buro Happold) under construction. The workshop gallery with green wood furniture.

Above, top: Munich aviary view of the mesh and mast head.

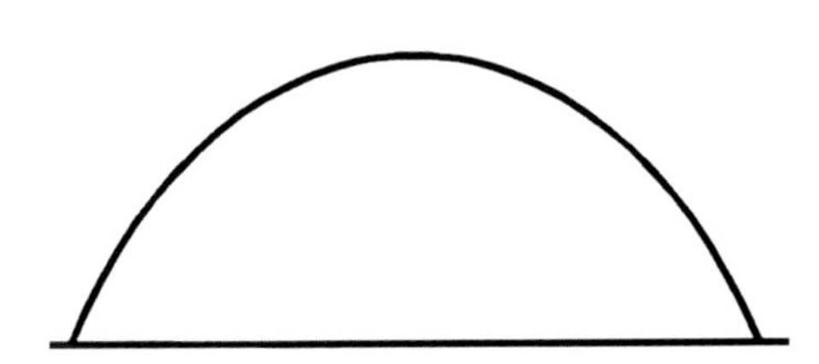

Ted invited John Howell, an aerodynamicist, who had worked with Ted in Structures 3 on the Mannheim timber gridshell, back from Canada. He brought in Chris Williams, also from Structures 3 and the Mannheim project, for the computer modelling; for the model making and testing he brought in Dennis Hector, an American architect whom Ted had known at Frei Otto's Institute for Lightweight Structures in Stuttgart. And he collected together several materials scientists, including Bryan Harris who was already at Bath, as well as both structural and environmental engineers.

In 1979 the team was lucky enough to acquire a full-size air-house that someone had paid for but not used, and erected it on a car park on the university campus. Using this full-size structure the team was able to make the all-important comparisons with results obtained using small-scale physical models and computer models in order to calibrate their experimental models and turn them into useful design tools. Their work embraced the behaviour of the membrane material and the entire structure under wind and snow loads, the influence of air pressure as well as the environmental characteristics of the structure – the acoustics, internal air movement and thermal behaviour.

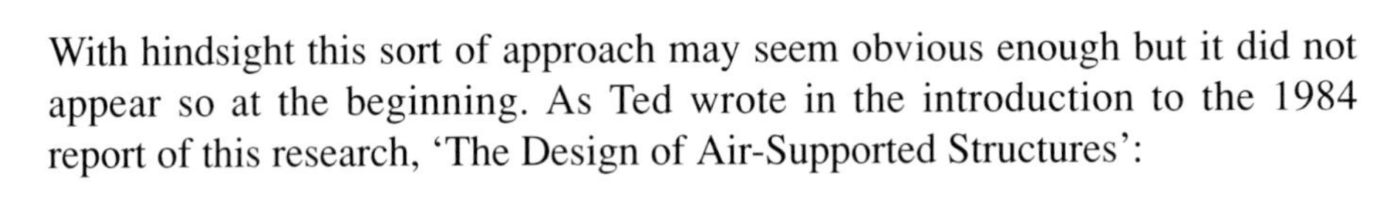

With hindsight this sort of approach may seem obvious enough but it did not appear so at the beginning. As Ted wrote in the introduction to the 1984 report of this research, 'The Design of Air-Supported Structures':

> When I first arrived at the University of Bath in 1976, it seemed right to try to set off in a field of research in the same way as one approaches design in that I should try and gather a group of friends with different interests and skills and together we should take one type of structure and examine as many aspects of it as we could in order to provide a basis for integrated design methods ... Perhaps the problem with university research in this country is that it encourages many small pieces of individual study, yet there is little attempt to relate them together so that their relevance is tested and so that they can be used in practice. Here we are still attempting to encourage individual pieces of research, by both people in academic life and in practice and, additionally, to stimulate discussion and use in practice so that we can uncover the true areas of concern in order to eliminate the irrelevant.

Perhaps Ted's greatest achievement was attaining eminence as an engineer in the academic world and in the world of professional bodies. This was no better exemplified than in the engineering research work he did or inspired others to do. At the time, the air-house research was exceptionally rare in that it united the efforts of university staff, consulting engineers and the professional body, the Institution of Structural Engineers. As the 1990s draw to a close, such collaboration is becoming more common.

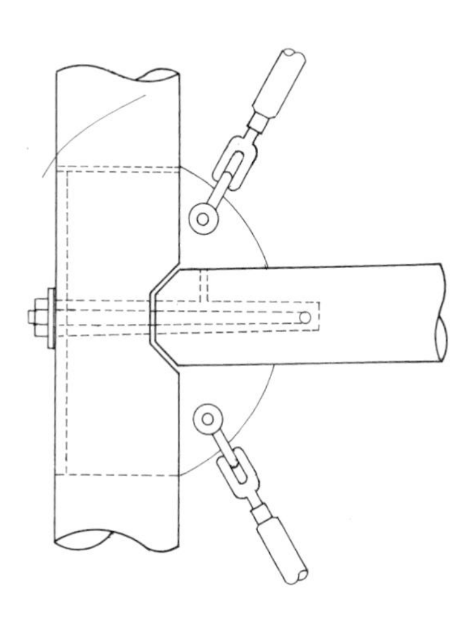

At about the time the air-house research was finishing Ted Happold took a similar approach to an aspect of building design and construction that he felt intuitively would become more important. It had taken a world oil crisis in the 1970s to awaken people's concern for our civilisation's sources of energy, and resources in general, although, even today, it cannot be said that the message has been embraced by everyone. This concern has led to a dramatic change in how good buildings are designed. As well as minimising the amount of material used – long of concern to the structural engineer – it has become equally important to minimise the amount of energy a building uses during its life, and even during its construction. Ted was relatively early among engineers to realise the significance of this for the profession. In particular he understood the vital role the façade plays in environmental engineering, and the many problems its successful execution raises, not least of which is the question of final responsibility for its performance.

Like many of his achievements, Ted's plan to set up a specialist research and development unit was based on ideas he had seen elsewhere. In 1983 he had visited a small research institute for the building envelope at Rosenheim in Germany and at once recognised its value. Ted's particular skill was transforming this idea into something that would work – in this case in a university in the UK context. With the collaboration of his old firm, Arups, and a broad selection of specialists in the industry he generated the enthusiasm to set up the Centre for Window and Cladding Technology and used his influence at the University at Bath to find it a home. It has now become one of the world's leading institutes in this field, embracing, the entire range of issues and disciplines related to the envelope and dedicated to generating results of immediate use to building designers and constructors.

Opposite, top left: Close up of the greenwood components, Hooke Park. Centre, left: Air House section. Bottom left: The Wild West solidity of the post and platform work. Top, right: The Research Team on the Air House, University of Bath.

Above, left: within the shells of the completed training centre; top right: model of Hooke Park Masterplan (ABK); bottom: section through the rigging joint.

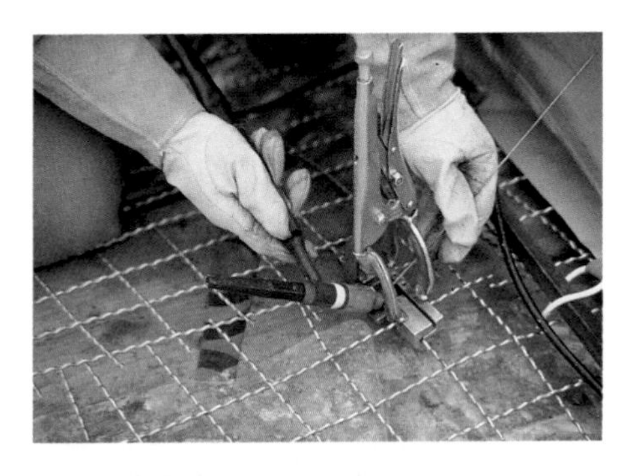

During this same period Ted was engaged in another project, through the Institution of Structural Engineers, which he considered to be research, although the university community might not acknowledge it as such. Rather earlier than many engineers, Ted realised in the late 1970s that an increasing proportion of the engineer's work would be the appraisal and rehabilitation of the growing stock of existing buildings. He formed and chaired the Institution's ad hoc committee on The Appraisal of Existing Structures. The report of this Committee in 1980 became a classic at a time when there was very little guidance for engineers in this aspect of their work, with a result that many old buildings had been needlessly demolished in the 1960s and 1970s simply because too few engineers were able to argue their viability. During the 1990s Ted chaired the Institution's Task Group which completed the second edition of the work a few months after he died.

Despite his skills in working with academic and professional institutions, Ted most enjoyed live projects and hands-on research. He loved making structural models and testing them to see how they worked and how they could work better – research with almost boyish enthusiasm, and a passion he shared with Frei Otto. After his move to Bath, Ted was closely involved in two projects that were archetypal, both in being collaborations with Frei and using a combination of research and engineering analysis to develop outline design concepts to achieve unique and elegant simplicity. These were the aviary at Munich Zoo and the several buildings at Hooke Park for the furniture designer John Makepeace which are made from forest thinnings, timber which normally would be burnt as firewood.

Munich Aviary

The aviary at Munich had to be, in the words of the architect Jorg Gribl, 'a structure like a cloud' to cover 4,500 square metres. Making clouds was just the sort of challenge to appeal to Ted. It is customary to keep birds in an aviary using a steel mesh with holes large enough to allow small birds and mammals, but small enough to retain larger birds. It is not uncommon to suspend such a lightweight, unstressed mesh from a prestressed cable net structure, made from large cables spaced a metre or so apart. But it seemed wasteful to provide two nets. Why not instead make a prestressed structure, similar in form to the membrane tents that the Buro Happold team had built many times with Frei Otto, but using a high strength, fine-wire steel mesh?

To understand the nature of this problem it needs to be understood how a flat woven mesh can deform to create a surface that curves in three dimensions. This occurs by the weft and warp shearing relative to one another so that the initially square holes in the mesh become diamond shaped. If this shearing cannot happen freely the mesh will crinkle – just as a tightly woven fabric does over curved parts of the human body, while a loosely woven one hugs the three-dimensional shape. However, a mesh woven from wires large enough to carry the stresses would be too resistant to shearing and vulnerable to the wires sliding past each other, as well as being very costly.

As with most original design, the solution seems obvious with hindsight. It presented itself in the form of some rolls of fencing that Ted Happold and Michael Dickson noticed while on a visit to the zoo. They were woven from crimped stainless steel rods 3.2mm in diameter, and available as a standard product in rolls 50 metres long and 2.5 metres wide. The crimping allowed easy rotation of the weft and warp while preventing the wires sliding past each other. This was the birth of a solution, but needed a research programme to turn an idea-in-principle into something that would work and could be built. The corrugated mesh was, effectively, a new construction material whose properties needed to be measured and for which methods of off-site and on-site fabrication and a means of erection had to be devised. How strong were the corrugated wires? How easily and by how much could they shear? How could roles of the mesh be welded together? How could the tension in the mesh be concentrated to the points of attachment at masts and tie-downs without overstressing the wires of the mesh? How could the entire, pre-assembled mesh be lifted into position?

The research was carried out in three locations – the strength and stiffness of the wires, the welding technology and the attachment to masts at Bath, the form-finding with Frei Otto in Stuttgart, and the analysis of stresses in the mesh and the cutting patterns with Mike Barnes, then at City University, London, and now Ted's successor at Bath. The result speaks for itself.

Hooke Park

John Makepeace had approached Ted Happold to help devise a plan to transform 330 acres of low-value woodland by forming a school for woodland industries at Hooke Park. Makepeace approached Richard Burton of ABK Architects to prepare a master plan, with Buro Happold as engineers. An essential characteristic of the buildings was that they should be created from products of the wood itself and achieve a consistent architectural and engineering aesthetic. The first two buildings, the prototype house and the training centre, were designed by ABK working with Frei Otto; the third building, the Westminster lodge, was the last of Ted Happold's several collaborations with Ted Cullinan.

One of the keys to the success of the projects was the development of a simple joint that could be used for green timber. The conventional method of fixing a steel rod into a longitudinal hole with epoxy resin would not achieve sufficient strength in tension – the glue would be stronger than the bond between adjacent fibres and the rod could easily pull out. Using the research labs at Bath, Ted helped develop an ingenious solution. If a stepped hole were made, some of the glue could be forced into the fibres themselves which would allow them to carry tensile stresses (which are more effectively than shear stresses). The result was a more effective load transfer between timber and steel and a joint which could carry the required tensile loads.

Although hidden from view this is a wonderful example of developing the appropriate engineering aesthetic for a new material. Each of the three buildings on the site is clearly an engineered building and incorporates forms and construction details that are unique to the roundwood raw material – the various connections, the sharp curves that can be achieved in green timber, and the versatility that allows it to carry compression, tension and bending loads. Without the design and research input from consulting engineers, none of the buildings could have transcended the construction vocabulary inherent in craft-based carpentry skills.

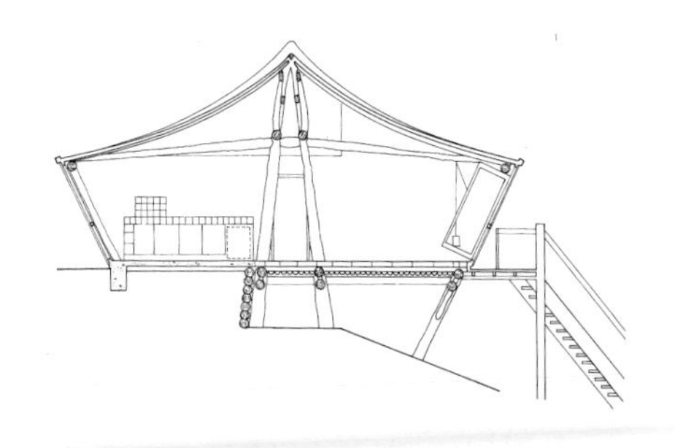

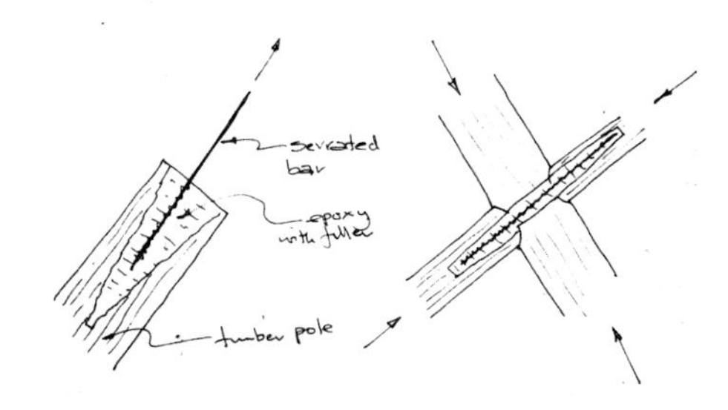

Opposite, from top: Crimping the mesh, Munich Aviary; Close up of the crimped stainless steel 3.2mm diameter mesh; Mast head and pantograph system of cables linked with mesh clamps collects the load from every mesh filament around the perimeter of a roughly circular hole in the net and concentrates it into just four adjustable cables attached to the mast head; Modelling of the mesh, Michael Dickson to the side.

Above, from top: Staff house section (ABK, Frei Otto, Buro Happold); Middle: Staff house under construction; Bottom: Ted Happold's drawing of joining systems for round wood timber.

Recent Projects

Edward Cullinan and Partners

It was about seven o'clock in the evening; I was in the office drawing and dreaming, first glass of wine almost empty. The phone rang, I picked it up; the phone said 'Hello, hello, hi Ted, Ted; I'm at Heathrow, off to see Frotto. Had to ring, brilliant news, at last we're going to work together, best job ever, brilliant, can't wait, must go, plane's leaving.' By then I'd realised that 'Ted, Ted', meant 'Ted, this is Ted Happold' and the brilliant job meant he'd got the engineering at our job for MEPC/IBM at Bedfont Lakes. I told him I was dead pleased too and off he flew. I didn't put it together until later that Frotto meant Frei Otto who was co-engineer with Ted on that incredibly simple, inventive woodworking shop, the most elegant flowing interior for working wood there is, by Richard Burton at Hooke Park in Dorset.

Ted could work with anyone; he was infinitely outspoken, inventive and optimistic, but gentle with it. At Bedfont he and his firm made important contributions to the buildings by us and by Hopkins; but his crowning glory is below ground in the beautifully designed, elegantly lit-by-lights-shaft, three-storey car park. It needs visiting in its own right. With him we did the greenwood house at Hooke Park and he was endlessly inventive; he and his partners invented with us the lovely bridge at Warwick, and with his partners we are doing the Faculty of Mathematics at Cambridge. I have never worked with a more contributive, inventive, bright and joyful engineer, or any other kind of consultant for that matter: and he's passed it on. Just before he died, we were having a drink together in his lovely home in Bath, when he said of himself, 'The structure's fine, it's the pump's the problem,'. He was wrong, he left his heart to all of us. TC

Opposite, clockwise from top left: Pedestrian bridge, International Manufacturing Centre, Warwick University (the bridge spans 43m and the main structural element is a tubular steel truss 43m long) Hooke Park, where Cullinan succeeded ABK working with Buro Happold on additional buildings; The first building Westminster Lodge, shown in detail during construction; One of Cullinan's buildings for New Square Bedfont Lakes set within the guidelines of Hopkins' master plan; 18m-wide strip buildings ending in complex tripartite gables (this feature is repeated in the central portico of each block).

Above, from top: The two Teds, Cullinan and Happold, in deep conversation shortly before Ted's death; A detail of the pedestrian bridge at Warwick University.

Above: Multi-level linking walkways within the atrium. Use of a stretched fabric structure gave greater freedom to cover the space with minimum support structure necessary to accommodate the complex geometry. The roof has very low self-weight.

Imagination, Store Street, London

Completed in August 1989, this extremely rigorous £5 million refurbishment was achieved in one year. It provided Ron Herron with precisely the vehicle and proactive client to produce the minor masterpiece that his career had always suggested but never accomplished. Buro Happold acted as structural, fire and services engineer for the project.

The initial prognosis was not promising: two parallel buildings, reconstructed after the Second World War, with an unused and extremely bleak brick-faced gap between them which absorbed most of the light penetrating the space. The tour de force was the bridging of the gap with a geometrically complicated roof covering which magically transformed the interior and provided a circulation hub. The building dramatically accentuated by the skeletal steel staircases and bridges which linked the flexible open-plan spaces required by a high-tech company for whom image is of prime importance. The roof carries on the commitment Buro Happold have made since their inception to high-quality lightweight structures. The accurately patterned membrane of fluropolymer-coated PVC polyester material is divided into fourteen panels approximately 7.5m x 5m. Each membrane panel is supported on a domed top flying post of stainless steel wires spanned between light bowstring steel and aluminium support beams. This design clearly expresses the advantage of translucency offered by such materials in providing enclosure to an intermediate environment. Added to this is the extreme lightness of construction and ease of erection that has been made possible using techniques of accurate prefabrication and a minimum weight structure.

Right from top: The atrium at night. To reduce the level of force on the existing buildings a lattice supporting structure was developed across the building. The whole roof is enclosed in the transverse direction to the front block; Form of the atrium roof covering showing paired masts

The Royal Armouries, Leeds

It is perhaps appropriate that the last building Ted Happold and his practice should realise before his death is the Royal Armouries Museum. Ted joins that other remarkable Leeds engineer John Smeaton in the pantheon of great engineers with a contribution to the city he loved which will honour his name and help consolidate his reputation.D.W.

Not surprisingly the profession of military engineering came first with the construction of castles and fortifications. Nowadays, as the Gulf War and Bosnia show, it is more devoted to machinery. But civil and building engineering did not take long to follow with the development of the great cathedrals of medieval times. Of course cathedral building was more than engineering. It was an art-form raising emotion through soaring space and the enclosure of light in the service of the development of religious symbolism. This was architecture.

Architectural possibilities, however, are rooted in engineering invention. Engineering is about organising the design, manufacture, construction, operation and maintenance of any artefact which transforms the physical world around us to meet a recognised need. It provides an effective service to architecture because it is about economy. The engineer surveys the site for information on the soil beneath, its exposure, its environment. Here the site is on the side of the old ford across the River Aire, subsequently where the Leeds–Liverpool Canal and the Air and Calder Navigation ended at Clarence Dock. The site is subject to flooding and where once water would have been used to keep people out, now a fortification is needed to keep the water out. A defence wall has been erected right round the museum building on the river side, and downstream the ground level of the Tiltyard is raised to 300mm above the 1949 flood level.

While the outside of the building has to be defended against the river, the inside contains a great number of very valuable exhibits susceptible to deterioration. This calls for a very stable internal level of temperature, air movement and humidity. Light is necessary but is also corrosive to exhibits. A satisfactory environment is achieved by a massive envelope of blockwork and concrete exposed so that it reacts directly with the space, punctured by very small windows in order to minimise solar gain. Because evenness is so important, the building had to be air conditioned to guarantee ventilation. The system selected was a displacement air-ventilation system. Cool air is introduced into the gallery spaces at low level and at low velocity. This is then mixed with the ambient air by the movement of people in the space; as the air warms – from the heat of lighting, other equipment, people and so on – it rises through the gallery space and is extracted at high level in the ceiling above.

For aesthetic reasons the building services and ductwork are unobtrusive yet easily maintained. Some of the exhibits are very heavy and so the floors to the galleries have to be very strong. The main floors have twin beams and between them 'trenches' in the floor to take the ducts providing air and below a ceiling void for the extract air from the ceilings below and exposed waffle slabs to make the floors between the beams and ducts. A similar approach has been used to other areas.

The mass of the building contrasts with the tower of the Hall of Steel, which with its light structure and elegant planar glazing as an effective counterpoint. Interestingly enough, the end result is reminiscent of where the exhibits came from – the Tower of London's White Tower, with its ditch, 1.5m-thick walls and very small slit windows. An example of the advantages of historic military engineering!

Ted and Rod Macdonald in *The Making of a Museum*. Derek Walker, 1996)

Opposite: The museum from the River Aire.

Above, from top: Model prepared with the master plan for Clarence Dock showing the museum top right overlooking the river and the dock; Some 3,100 objects are displayed in the Hall of Steel; View looking up to the pedestrian bridge.

Page 142 top: Laubin perspective of the museum as built; Bottom: The Tournament Gallery.

Clockwise from top left: The Oriental Gallery; The Japanese Gallery; Main staircase Hall of Steel; Laubin cross-section showing main galleries and relationship to the dock; South entrance to street.

Foster Associates

Most of Norman Foster's considerable body of work since the early seventies has been engineered by either Tony Hunt or Ove Arup and Partners. It was not until 1986 that Buro Happold was appointed, initially on a tiny project in South Kensington for Katharine Hamnett.

The site was a dilapidated triangular-shaped two storey car repair shop connected to the Brompton Road only by a long passageway. It seemed at first an unlikely setting for a high-fashion clothes shop. However, in the eleven-week contract period asbestos was found and removed, existing external walls and the first floor were completely stripped out, internal strengthening steelwork was erected, the roof recovered, a new concrete ground floor laid and the wall adjoining the underground strengthened and repaired. Finally, new mechanical and electrical services were installed and the shop completely fitted out. The 'tour de force' was the installation of a glass bridge linking street and showroom. The design of the glass and its immediate supporting structure was fabricated to meet requirements varying from servicing to pointed high-heel shoes. The 25mm-thick glass with finely ground and profiled edges had a load capacity of approximately $20kN/m^2$.

To turn a dilapidated car repair shop into a spectacular fashion outlet required an extraordinary amount of energy, co-ordination and co-operation between consultants. This factor was not lost on Foster and though Buro Happold and Foster were on different teams for the Commerzbank, Frankfurt – which Foster eventually won – he could not fail to admire the quality of engineering implicit in the Ingenhoven scheme which was runner up. It is not surprising that as a result of these close encounters Foster used two of Happolds' strongest cards – their Islamic experience for the Al Faisaliah office complex in Riyadh and their pre-eminence in lightweight structures – to mount a successful competition challenge for the Great Court at the British Museum.

Company Headquarters Tower in Essen

After working together on the second prize-winning design for the Commerzbank in Frankfurt won by Sir Norman Foster and Ove Arup and Partners, it seemed inevitable that national interest in the ingenious pursuit of a green building – personified in the Frankfurt solution by Christoph Ingenhoven, his consultant Frei Otto and Buro Happold – would bear fruit. The team were fortunate to be retained and commissioned to develop their environmental solutions for a major office tower in Essen. Though Happolds were not permitted to implement their proposals as site engineers, it is apparent that the project changed little from the detailed design proposals they had prepared.

The requirements of the brief for an energy-saving design are reflected in the form and construction of this tower block. In terms of the ratio between the area of the outer skin and the volume, as well as aspects such as wind pressure, heat losses, structural cost and daylighting, the cylindrical shape represents an optimum form. Despite the continuous glazed skin, the different uses – entrance, office storeys, service floors and roof garden – remain legible. Vertical access is via a separate but linked lift tower, which also assisted orientation on the individual floors. The working storeys are laid out with a core zone for group and communal uses (sometimes with linking staircases between levels), a circular access corridor and an outer zone of offices. The form of the floor slabs, which taper towards their outer edges, maximises the penetration of daylight. Clerestory windows allow light to enter the core zone. The windows in the inner skin of the two-layer glazed façade can be opened, permitting all work places to be naturally ventilated. At the top of the tower the façade is also extended to protect the roof garden against strong winds. The building is designed to allow users to control their own environment. When weather conditions prevent the opening of the windows, ventilation is provided by an air-conditioning plant with a minimum air-change capacity. The concrete floors are clad with perforated metal sheets to enable their mass to be exploited for thermal storage.

The two-layer ventilating façade consists of an outer glazed skin on a conventional, thermally insulated exterior. Between these two layers is a 50cm-wide 'corridor', divided off on every floor and on every axis. The sunscreening system housed in this corridor is protected against the weather and has a positive influence on the energy balance since it restricts the heating-up of the intermediate spaces. The inner façade consists of room-height sliding elements that can be opened to a width of 135mm. The external skin consists of 2m x 3.6m panes of toughened safety glass with point fixings. Fresh air enters the convex-shaped façade elements through 150mm-wide slits and over sheet metal deflectors. The outer metal members of the façade elements are perforated near the floor on the right-hand side and unperforated on the left. This layout is reversed at the top near the ceiling so that diagonal through-ventilation is guaranteed and there is no danger of re-entry of stale air. All inner and outer façade components are integrated into prefabricated elements one-bay wide and one-floor high.

Opposite, top: Interior of Katharine Hamnett showroom, South Kensington (Foster Associates, Buro Happold); Glass bridge and entry way.

Above, top: Commerzbank, Frankfurt, as built (Foster Associates with Ove Arup and Partners); Bottom: Commerzbank Frankfurt – competition entry by Christoph Ingenhoven and Buro Happold.

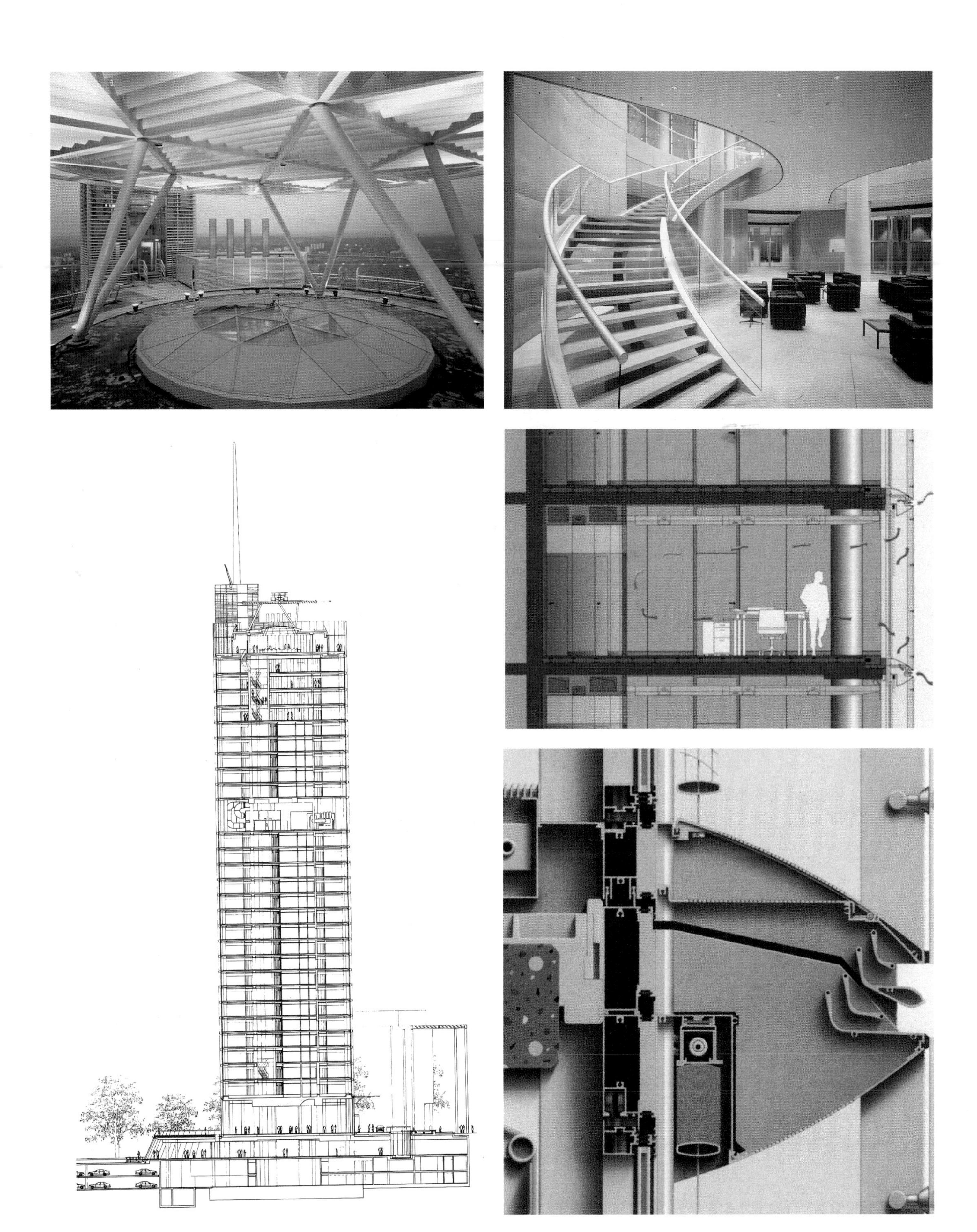

Top: Two internal views at roof and foyer level; Bottom: The main cross-section and details showing the ingenious façade systems developed for the building; Opposite: Four views of the Essen Tower (Christoph Ingenhoven, Buro Happold).

The Ones That Got Away

Vauxhall Cross

In the winter of 1981/82 a two-stage architectural competition was launched for the redevelopment of land at Vauxhall Cross on the south bank of the River Thames. The mixed development was to comprise one million square feet of office space, 240 luxury apartments, retail and community, and a public park and promenade along the riverside. Kit Allsopp and Andrew Sebire formed a team with Ted Happold for the competition, which attracted 128 entries for what was then a huge project with a construction cost of £200 million. In essence the scheme was a simple yet articulate integration of the main ingredients of the brief. The offices form a linear zigzag along the back of the site, maximising frontage and giving superb views up and down the river. From this wall of offices the apartments, suspended above a riverside park and connected to a raised and glazed shopping galleria running the length of the development, cascade down to the riverside.

This project once again demonstrates Happold's deductive abilities as an engineer. Because the Victoria tube line runs under the site, only limited opportunities were available for the foundations of the principal office block. Borrowing from his experience in Hong Kong, Happold introduced a limited number of high-capacity hand-dug caissons, carefully positioned to co-ordinate with the plan of the zigzag block. This he structured with light (10 kN/m^2) steel metal decking with its concrete topping hung off columns on the caissons with a chevron array of stay cables. A most innovative and economic solution. Sadly, following the granting of the Special Development Order by Parliament the developer Arunbridge went into liquidation and the project was never realised.

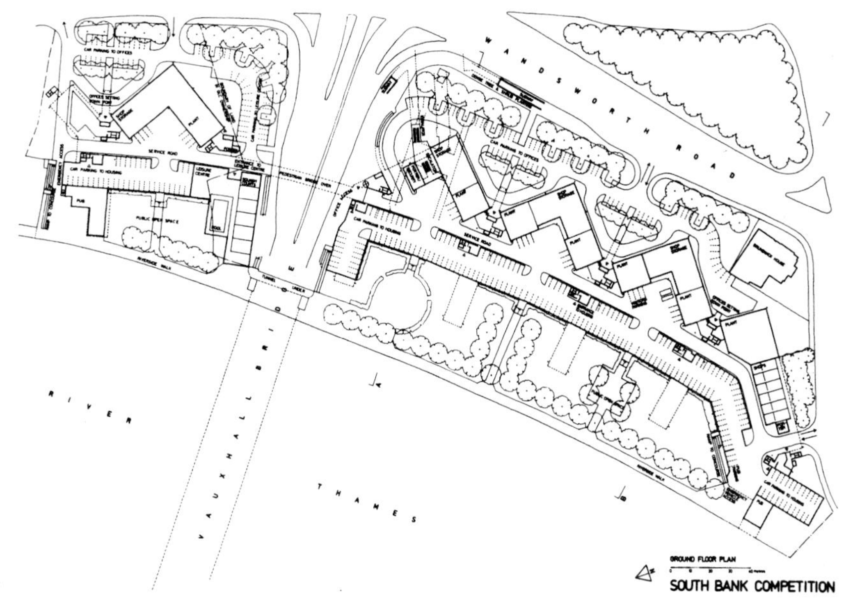

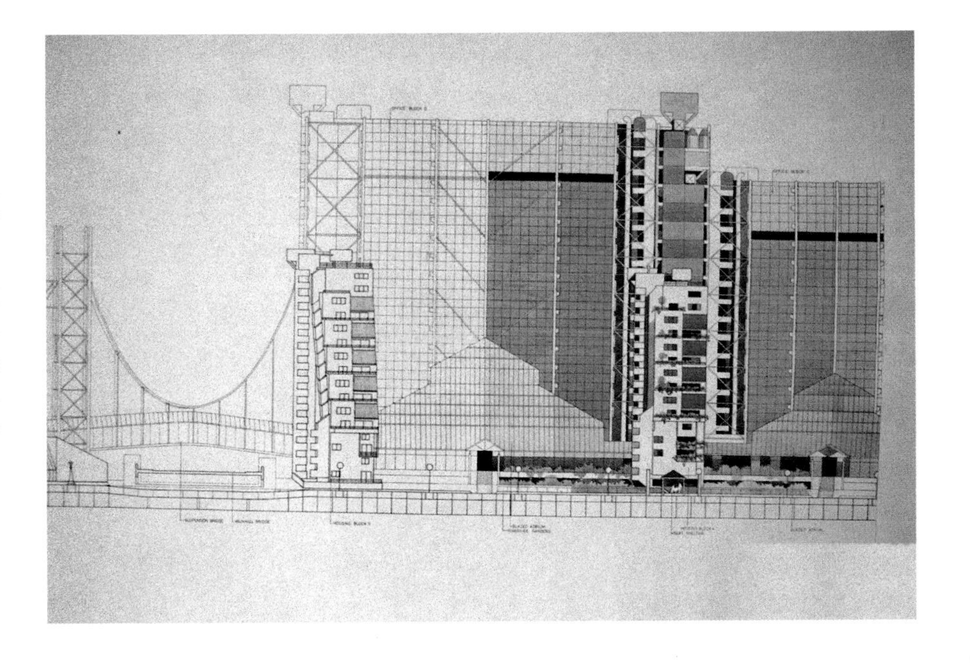

Opposite, from top: Elevation Vauxhall Cross; Sags contemporary cartoon; Section of the competition entry;
Right from the top: Model of the Project Site Plan and Detail of the Competition Project.

Arctic City, 58°N

In 1980 Buro Happold were selected to join Canadian architect Arni Fullerton, Dennis Wilkinson and Frei Otto to prepare a feasibility study for a covered township on the Athabasca river in northern Alberta. This project was generated by the planned expansion of tar sands extraction – later abandoned when the oil price fell in 1982.

Two structural solutions were proposed by the team. One was a large cable net mast-supported structure which covered about 4 acres. The other was a 35-acre air supported roof structure. The study ran into five volumes covering all aspects concerned with developing a new township, including the psychological problems of working communities in extreme northern climates, the physiological effects of living in an environment cut off from natural light and air, the technical problems of economic environmental control in an adverse climate, fire and smoke control within an enclosed space, and the actual construction of such a space. The study concluded that a large 35-acre air supported roof was best able to provide the suitable 'free field environment'. The roof was to be as transparent as possible, with a clear view to the horizon on all sides. The supporting structure would be free from visual noise. It was estimated that the enclosure would in effect move the internal micro climate of the township 10° of latitude south.

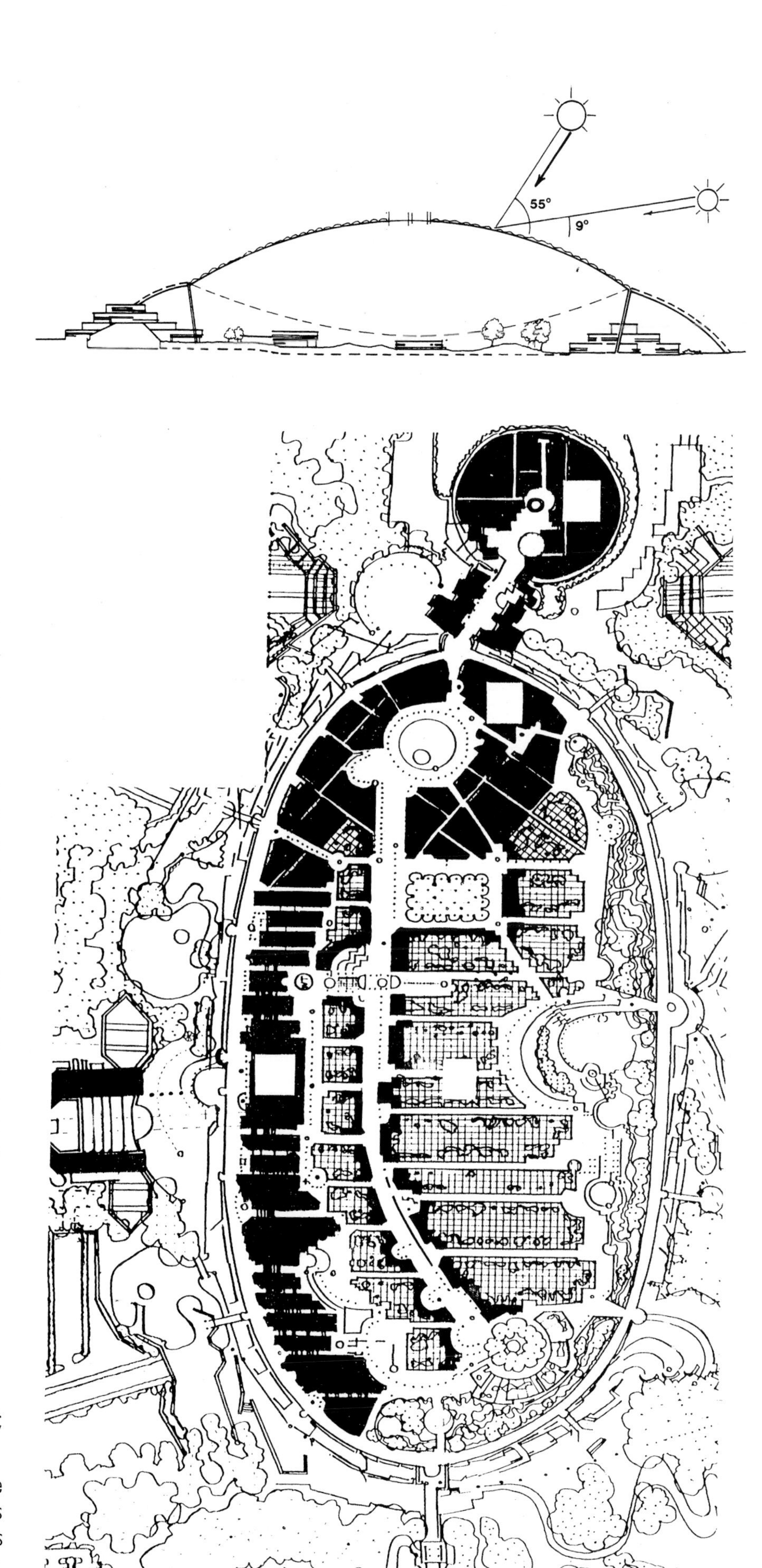

Opposite: Arctic City, 58ºN; Top: Photograph of the model; Bottom: Plan view of the model.

Right, from top: Section proposed; NS section through the enclosure; Plan of township. The diversity of requirements tested Buro Happold's diverse skills to the full.Contributions were made by Ted Happold, Ian Liddell, Roger Webster, Michael Dickson, Dennis Croome, Gareth Jones, Mick Green, David Wakefield, Peter Moseley and Mike Cook.

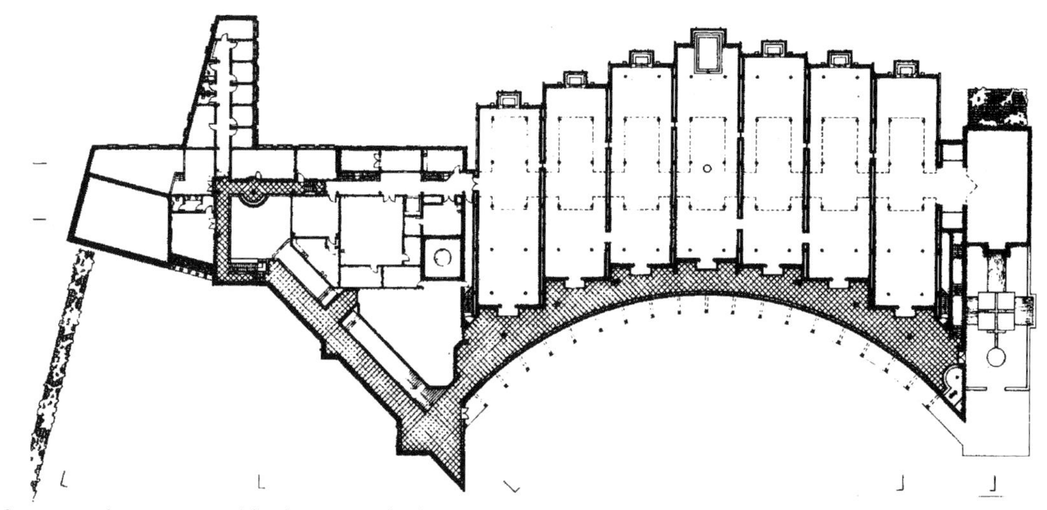

Above: Plan of museum; Below: Museum model and perspective sketch

Michael Brawne: The National Museum of Archaeology, Amman, Jordan

This new museum was to be located at the top of the eastern slope of the upper citadel in a partly excavated area adjacent to the existing museum. The new building then places its back to the prevailing summer wind. Extensive irrigation requirements for new planting were to be met by a 500-square-metre reservoir beneath the new museum. The style and nature of the new construction was to be evocative of other ancient stone constructions on the site and the museum was to be built in two phases. Phase One consists of galleries at the upper level, their undercroft and plant rooms. Phase Two contains the walled entrance courtyard with water features, reception, library and lecture theatre at the lower level and VIP rooms, administration offices, exhibition and conservation rooms at the upper level. The structure is a simple reinforced concrete frame system offering the possibility of a modulated façade of stonework stabilised against various loadings. The flexibility of the internal gallery spaces and their tempered environment is achieved by a network of plastic pipes placed in the floor screed below the marble floor finish. In winter this coil of piping conducts low temperature hot water, and in summer cool water from the reservoir below.

Sculpture Park, Milton Keynes

In 1977 Derek Walker Associates were asked to give form to a proposal by the Milton Keynes Arts Foundation to create an International Sculpture Park in what is now Campbell Park in the centre of Milton Keynes. It was to provide studios for artists in residence, a variety of galleries and an education centre. It was supported by Henry Moore who offered major work to the park on permanent loan. Philip King and Anthony Caro also promised pieces for display in the galleries and outdoor settings. Ted Happold, who had recently moved to Bath, acted as civil and structural adviser to the project. The team's proposal consisted of a series of plateaued courts generated from a continuous ramp looking on one side to internal galleries of varying configurations and on the other to outdoor courts and glades. Drawing galleries and maquette rooms were provided at the transitional levels, and gathered around the entrance loggia was a library, education department, children's studio and gallery shop. The building was to be of masonry construction with structural glass components and membrane roofs over some of the open courts. The sloping site also offered the opportunity for a continuous series of stepped landscaped roofs providing promenade areas for outdoor sculpture and a processional route running parallel to the internal ramped walkway. The design was taken to a very advanced stage of development but fell victim to the savage cuts experienced by many arts projects in the late seventies.

Above and top left: Vignettes of route.

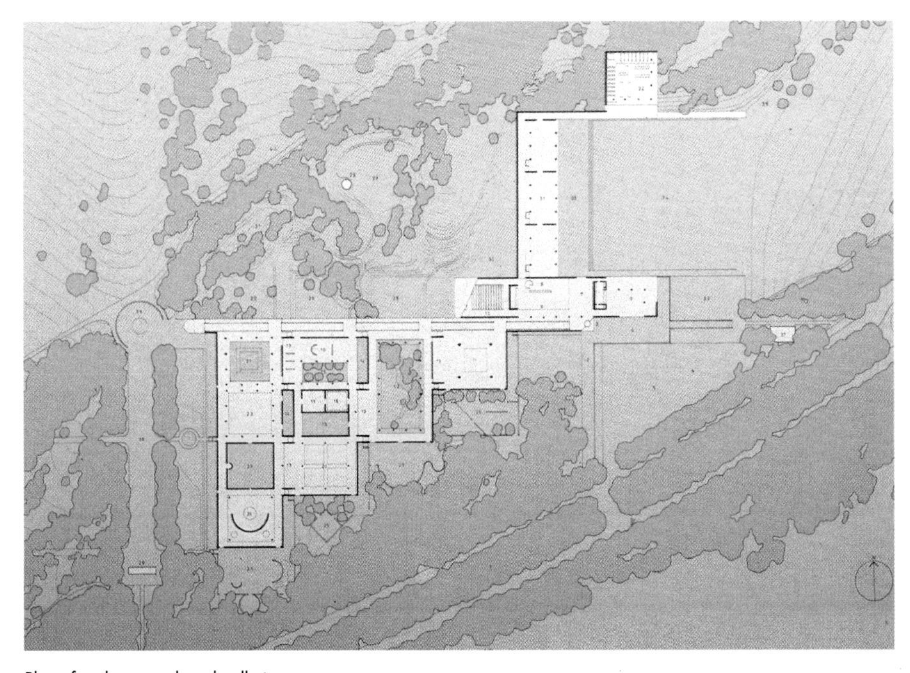

Plan of sculpture park and galleries

Pearls of Kuwait

As good quality coastal land had become a scarce commodity in Kuwait there had been increasing pressure to develop those areas of coastline which because of pollution and poor ground conditions had previously been regarded as unsuitable. Out of this was born the concept of new settlements built either on the coast around tidal creeks or in shallow waters just off shore.

Ted Happold and Terry Ealey were asked by the Kuwait Real Estate Company to undertake a feasibility study during 1987 to identify a number of potential development sites along the south side of Kuwait Bay and at Al Khiran, south of Kuwait City. To guide competing planning teams a summary report on coastal and tidal engineering aspects of the sites was prepared, together with a set of networks to creek and lagoon layout. After exhaustive evaluation, the joint winners were Philip Cox, Richardson Taylor and Partners of Australia and the Jerde Partnership of USA. The two schemes were significantly different in approach: Jerde's formal and broadly urban, Cox's more organic in form. In early 1990 a flagship project in Shuwaikh Bay was adopted and Buro Happold, Professor Patrick Holmes of Imperial College (Hydraulic Engineering), Anthony Farnes (Mariculture) and the Kuwait Institute for Scientific Research were commissioned to undertake the environmental assessment process.

The project was suspended following the invasion and occupation of Kuwait, and since the end of the Gulf War more pressing priorities have prevailed.

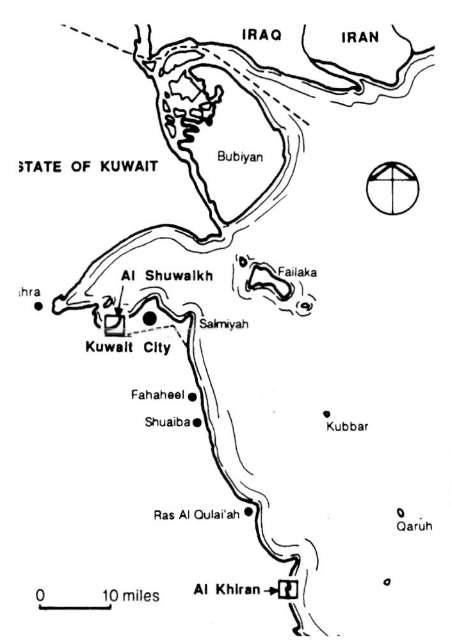

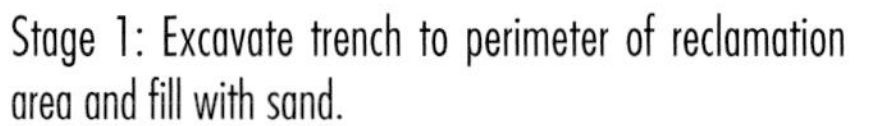

Stage 1: Excavate trench to perimeter of reclamation area and fill with sand.

Stage 2: Pump dredged sands in thin layer to start base of bund. Control slopes over silt to ensure stability.

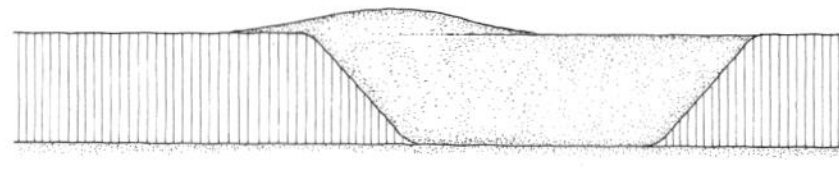

Stage 3: Build up bund in controlled fashion, ensuring mud waves are not excesive. Bund not likely to be built above high water because of shallow slope of reclaim when pumped fully above the water line.

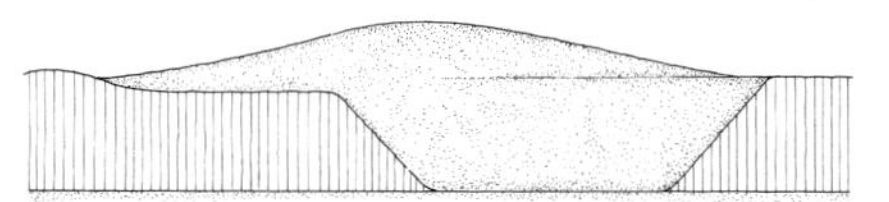

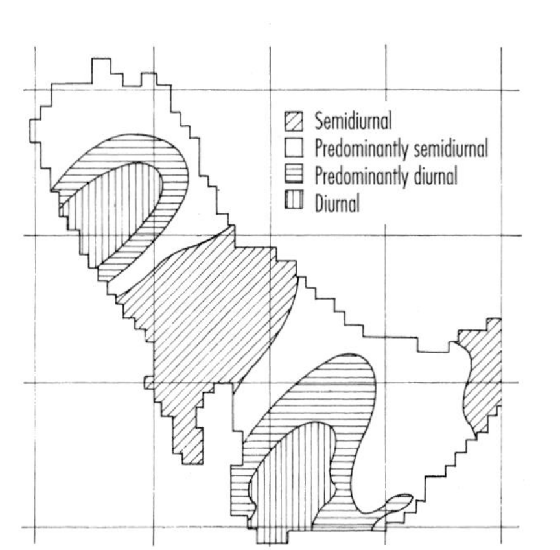

Opposite, left: Plan of Jerde project; Bottom: Perspective sketch of the recent project developed by Buro Happold.

Clockwise from top left: Site plan showing action areas in Kuwait; Sketch of Philip Cox's competition entry; Building of bunds diagram; The main Cox sketch for the original competition scheme; tidal movements in Kuwait.

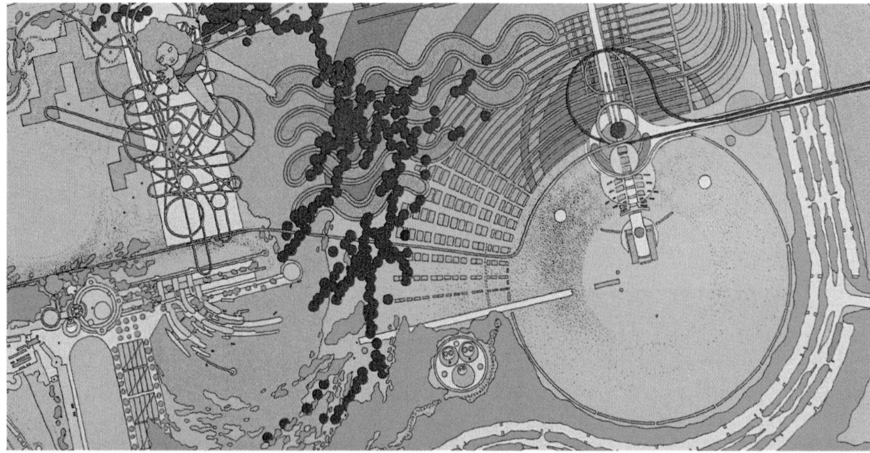

City Club, Milton Keynes and Energy Pavilion, Corby

Since the early seventies the design of large-scale leisure projects has been a preoccupation of Derek Walker Associates. The engineering implication of lightweight long span roofs and pneumatics formed a constant dialogue during this period between Ted Happold and Derek Walker; starting with Structures 3 and continuing with the new practice in Bath. The team produced a series of proposals for leisure environments and a city club for Milton Keynes during the mid seventies, all of which fell foul of the timidity and lack of imagination of both public and private sector operators. Nevertheless, they demonstrated a determination to find a way of combining the essential rationale of construction and engineering at their most sophisticated with a sense of playfulness, exuberance and a grasp of imagery and iconography of contemporary media and leisure culture. Many of the themes' stepped levels, orthogonal routes and lightweight roof structures were to be found in proposals for WonderWorld in the late eighties. This in turn led to another project currently being programmed for Energy World with construction based on a steel frame with insulated stainless steel cladding, lightweight steel floors and another Happold pre-occupation: the flat glass roof surrounding the geodesic domed Omnimax. A movable visor is powered by ambient solar energy and a computer-controlled array of parabolic mirrors provides solar energy sufficient to operate the 40m-diameter building drum. Externally working displays demonstrate renewable energy sources such as wind and water.

Opposite: Richard Horden's symbolic Wind Tower for Glasgow.

Left, from top: Study One for City Club, Milton Keynes; City Club for Milton Keynes (Bletchley brick pits); Energy Pavilion, WonderWorld Plc; Energy Pavilion interior.

Young Professionals

One of the most consistent and engaging aspects of Ted's personality was his devotion to and encouragement of young professionals. His position within the university led to many research initiatives that invariably involved young graduates. Within the practice the fellowship continued to expand, by very selective recruitment. His work at Arups had encouraged a number of young practices who now enjoy international reputations – Renzo Piano, Richard Rogers, Richard Burton, Paul Koralek and Jack Bonnington among many. Ted felt it was equally important to continue that policy at Buro Happold. Kit Allsop, Andrew Sebire, Trevor Denton, Richard Horden, Matthew Priestman and Christoph Ingenhoven have all benefited from close working relationships with the Buro. Happold's way has always been to respond to design challenges by harnessing collective knowledge and operating design seminars which operate on the same principles as a design studio in a graduate school. He considered that the subliminal selective memory which has experienced and assimilated most aspects of construction can develop a shorthand of options quickly and efficiently for the designer. Even when confined to his home by illness Ted still revelled in the intoxicating enjoyment of conceptual design meetings. His two-page summary outlining the 'Stuff of Life' reflected a series of projects incorporating his first love, design. Typically, they were wide ranging in scope, from small experimental works to major projects: a bridge over a motorway in Liverpool, a covered stadium for the Sydney Olympics, a mosque in Samarkand and a symbolic tower for Glasgow. One of Ted's most enduring qualities was his ability to connect during any design session – be it on a train, in his garden or in a more formal setting – with the critical issues of any problem. He had a sixth sense for spotting the route to an elegant solution.

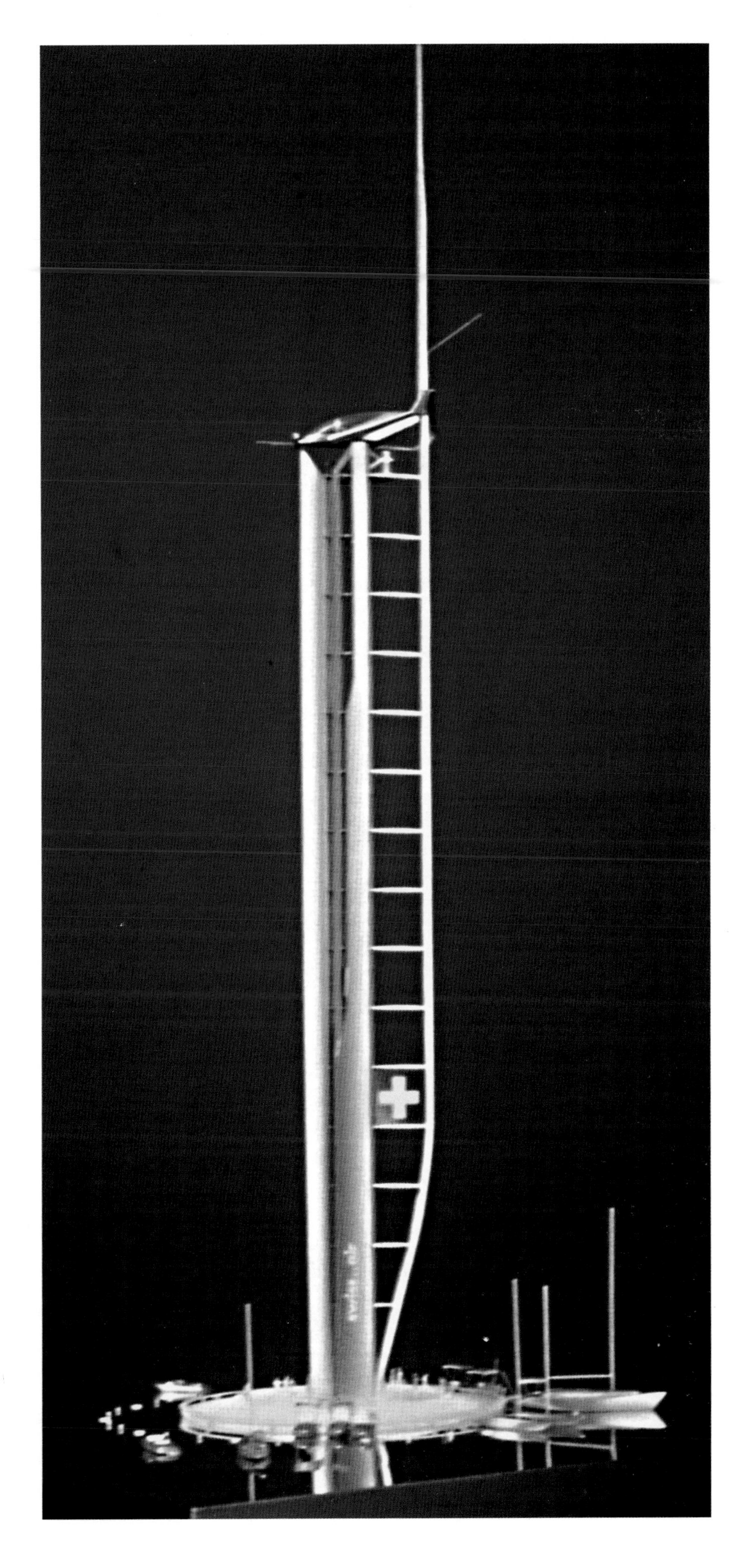

Legacy

It is interesting to examine the rich vein of projects that Buro Happold are currently involved in. It is perhaps a true reflection of the Happold doctrine. 'What I know about engineering is that it has to be a group activity, and the best work is done by the most diverse group of talents who can still live together.' 'G and C Rogers described engineering as the practice of organising the design construction and operation of any artefact which transforms the physical world around us in order to meet some recognised need. The product is usually the result of co-operative action on a large and certainly complex scale. Often the user desires to see the product or building designed by one person and it is certainly true that personal choice is an important element in most aspects of all design, yet the skills and experience needed in the whole process are rarely, if ever, held by one person.'

Ted Happold knew this better than anybody, and his team of original partners in Buro Happold, fellow engineers from Structures 3 and their later crop of associates and partners realised that collective responsibility and complimentary skills allow continuity and consistency that is the hallmark of great practices.

I read recently a brief article by Peter Dunican about Ove Arup that could have been written about Ted Happold. A journalist was having problems trying to pinpoint precise examples of Arup's personal oeuvre and after two hours could get no further than a footbridge in Durham. Surely, he said, that cannot be all he has done ... but this sort of question cannot be answered simply. It's rather like expecting the captains of football teams to score all the goals. Ove is a great individualist, but has always recognised the need for team working, for the intimate collaboration of equals in this most complex business of designing buildings. Ove established his pre-eminence through his firm and no further speculation about personal credits is necessary. Happold's legacy is mirrored in the respect and affection all his partners and staff felt for him. Their words need no embellishment.

Ted was a great starter of projects, a promoter of ideas, and a builder of teams. His generous nature and confident personality enabled him to gather round him a number of fine engineers and he inspired them to put in the extra effort to make their buildings successful. Five of these people are now still partners at Buro Happold. – Ian Liddell

Ted had enormous energy and he put that energy into achievements with his friends, in building, research, the Theatre Royal in Bath, public service and, sometimes, into working out a different way to travel from A to B because it would be more interesting and more fun. – Terry Ealey

The recent crop of competition successes and major commissions is a sure recognition that Ted Happold's ambitions for the practice have been realised – Eastleigh Tennis Club, Atlanta Olympic tents and the great Millennium Dome at Greenwich are further evidence of its supremacy in lightweight structures. The Lowry Centre, the British Museum Grand Court scheme and the Museum of British History are great national cultural initiatives and the prestigious office complex for British Airways at Prospect Park and the Al Faisaliah Centre in Riyadh extend the range of high-rise and environmentally sensitive structures that were initiated at Frankfurt and Essen with Christoph Ingenhoven and at Bedfont Lakes and Hooke Parke with Michael Hopkins, Richard Burton and Ted Cullinan.

Opposite, top: Hampshire Tennis and Health Centre at Eastleigh (designed by Euan Borland Associates and Buro Happold.) The structural costs were minimised by spanning two tennis courts with pre-tensioned cables supported by external support masts. Translucent ETFE foil cushions are supported from the cables. Overall stability is provided by a system of external ties and ground anchors; Bottom: AT&T tents for the Atlanta Olympics designed with FTL reinforced their position as the world's leading specialists in lightweight structures.

Above, from top: Internal view of one of the Atlanta Tents; Detail of the masts and the foil roof at Eastleigh.

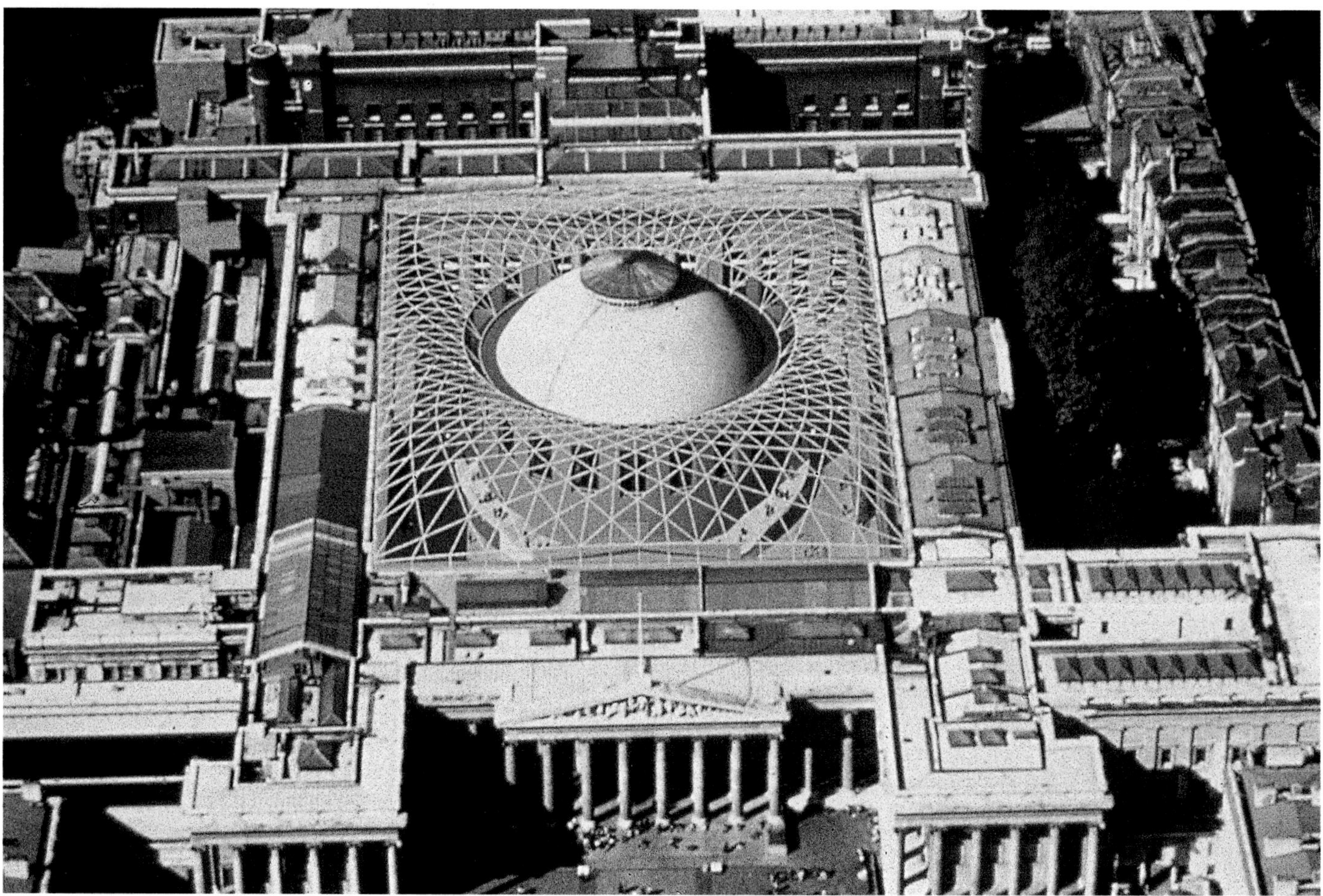

British Museum Redevelopment, London

The Great Court of the British Museum is one of London's lost spaces. It was originally intended to be the heart of the museum but almost immediately after it was built it was lost for public use by the insertion of the new Reading Room of the British Library and associated book stacks. This caused congestion in the front hall, and constricted circulation throughout the museum. The departure of the British Library provides a magnificent opportunity to recapture this central inner courtyard.

The Great Court is enclosed with a lightweight glazed roof. It would be entered from the principal level of the museum, through Smirke's great Ionic portico. The museum's Centre for Education and new ethnographic galleries will be located beneath this main level. Bookshops, restaurant and cafes will be on Level 2 and the mezzanines, elliptical in plan, are centred on the Reading Room. A pair of great staircases, which form a processional route linking the court to the upper galleries, encircle the restored drum of the Reading Room. The original Smirke façades of the courtyard will be restored and the southern portico reinstated. The new space, with its light-transmitting roof, will complement the nineteenth-century architecture of the museum.

The forecourt in front of the museum will be freed from cars and relandscaped to form a new external space complementary to the new public Great Court. Both will be open to the public from early in the morning to late at night, creating a major new public space for London.

Opposite, top: Cross-section through the competition entry proposed by Sir Norman Foster and Partners and Buro Happold for the British Museum; Bottom: The lightweight roof over the Smirke Courtyard at the British Museum.

Right: Computer generated images of the roof structure over the great drum of the Reading Room.

Millennium Dome, Greenwich

It is ironic, and sad, that the first formal collaboration since Beaubourg between the Richard Rogers organisation and members of the Buro Happold team should occur after the premature deaths of both Ted Happold and Peter Rice. Symbolically, however, it could not be a more appropriate project. The Millennium Dome will be the centrepiece of Britain's celebration of the new century. It will be a poignant reminder that Happold's long-term strategy for his practice has been fulfilled. Ian Liddell, Ted's longest-serving colleague, and Paul Westbury will lead a team of young engineers in the structural design of the world's largest dome, with another partner, Tony Macloughlin, leading the building services team.

The real challenge for the design team was in producing a concept for the structural support of a 400m-diameter umbrella that would have sufficient internal clear span to avoid restricting the flexibility of the exhibition that was still being designed below. It soon became clear that the most feasible solution was to use a tensioned cable net system that could then be clad with tensioned fabric to form a very lightweight and very cost-effective enclosure system that could be designed to take the form of the spherical cap.

After many weeks of development and design discussions, the primary structure of the dome was confirmed – seventy-two radial stringer cables (pairs of 32mm-diameter steel spiral strand) that were to define the 360m diameter building surface. These cables run from the building perimeter to a central cable ring that forms a hub, collecting all of the forces together at a height of 50m above the ground. Each stringer is individually supported along its length at a radial spacing of between 25m and 30m by an arrangement of upper hangers and lower tie-down cables that are arranged around the twelve 100m-tall primary steelwork masts. Circumferential cables are positioned above and below the surface to keep the stringers on their radial lines and connect into every node on the radial system. The radial cables are clad with a simple system of tensioned polymer-coated polyester membranes that are fabricated in tapered sections and connected into adjacent stringer lines.

The structure of the dome is entirely made up from tension-only elements: cables and membranes that can carry loads only through tension. It is rather ironic that the building has become known as the Dome as this traditionally refers to a form of construction that carries its loads primarily through compression forces.

The fabric and the radial cables are erected with sufficient tension to prestress them against imposed load deflections. As wind and snow apply loads to the building throughout its life, the structure will move until it forms a stressed geometry that can resist them. The prestress levels and the cable geometries have been selected through careful design and through analytical and physical model analysis to provide adequate deflection control, with the materials and structural sizes selected to provide very high stiffness.

The radial forces are collected at the perimeter by a series of twelve 75m-span boundary catenary cables. These 90mm-diameter cables skip around the building perimeter collecting the primary surface forces, transferring them to a series of twelve pairs of ground anchorages. By minimising the number of foundations, the forces are focused into the least possible number of locations.

Opposite: Computer generated image of the great Millennium Dome (Richard Rogers Partnership and Buro Happold).

Above, from top: Buro Happold staff working on development models; Development models of the ground anchorages; Computer generated image of the dome.

The Lowry Centre, Salford – Adaptable Theatre

The structure and environmental control of an adaptable theatre is one of the most interesting challenges given to the engineer-designer. This adaptable theatre is a small element of the Lowry Centre, designed by architects Michael Wilford and Partners with Buro Happold.

The Lowry Centre sits in Salford Docks on the promontory of pier 8 ,looking north, south and west across the canal contains a lyric theatre, the adaptable theatre, a gallery for the work of Lowry and a children's experience gallery for the visual and performing arts. The aesthetic appearance of this space is defined by the integration of the architecture and engineering. In an adaptable theatre the stage and auditorium operate in concert to provide a variety of settings which incorporate standard theatre requirements in totality and a variety of configurations and seating.

The balconies are suspended and cantilevered. The lowest balcony joins with the stall seating when it is in place and conceals it when it is moved away. The fire engineering strategy has allowed the steelwork of these balconies to remain exposed, expressing each detail and connection while still ensuring the safety of occupants. The balcony structures hang from deep steel trusses which also support the fly grid as it extends out beyond the formal stage. A double-layer floor and acoustic bearings separate the auditorium and the room above. Within the deck of the lowest balcony a plenum supplies cooled air through diffusers to both balcony and stalls seating. Directional nozzles from circular ducts provide cool air to the upper balconies. The efficiency of cooling for the whole project is improved by the use of canal water in the condensers of the main plant rooms.

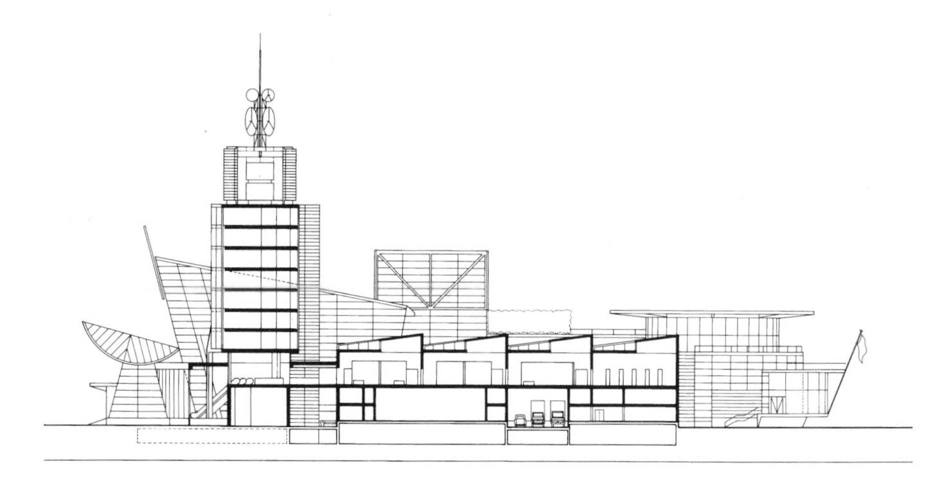

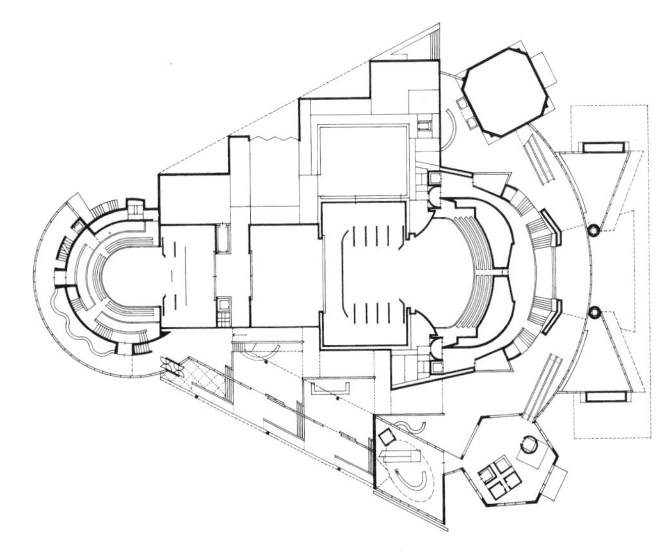

Perspective of the Lowry Centre on the promontory of Pier 7.

Above from top: Cross-section; Perspective view: Below: Floor plan auditorium level.

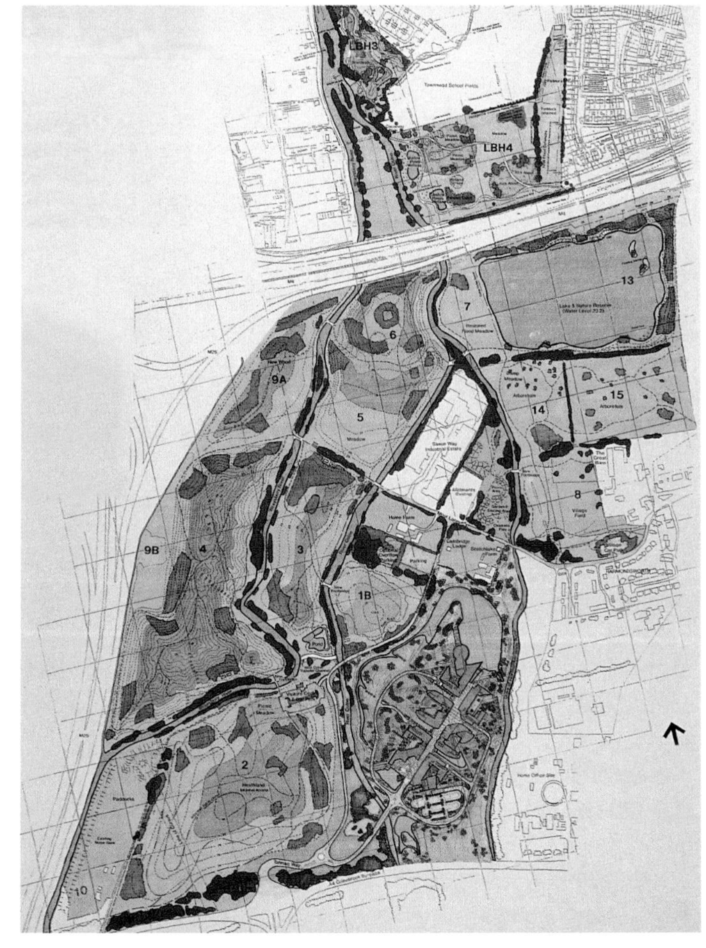
LBH3
LBH4
13
7
6
9A
5
14
15
8
9B
4
3
1B
2
10

British Airways at Harmondsworth, Prospect Park

Buro Happold were appointed as structural and civil engineers for the construction of a new Combined Business Centre for British Airways at Heathrow Airport in February 1993. The Norwegian architect Nils Torp had been appointed lead consultant after winning an international competition. His design incorporated six four-storey 'houses' joined by a covered street over a two-storey basement. The brief, developed with the design team, called for 50,000 square metres of office space and 1,850 underground parking spaces. The client was particularly concerned to produce a development with a human scale, encouraging interaction of individuals to promote synergy and generate a safe and friendly working environment.

The 200-acre site is situated to the north-west of the airport adjacent to the M4/M25 junction, within the greenbelt. Much of the site consisted of a number of worked gravel extraction pits, many of which had been used for landfill. The office development was to be on a 35-acre site, enclosed by two rivers, on the corner of the parkland. Much of the landfill was below the ground water level, which was supplied by the rivers crossing the site. To enable the removal of the fill it was necessary to first 'drain' the site. A bentonite slurry cut-off wall was installed around the full perimeter of the site to stop the river water recharging the ground water. The removal of the refuse gave the opportunity to incorporate a two-storey basement within the development and save the need to backfill that part of the site with imported material. In this respect the large basement proposed for the development resulted in a cost saving. In addition to the 1,850 parking spaces the basement was able to accommodate much of the air handling and boiler plant for the development and a service access route.

Above ground, the six houses each consisted of four floors of two 'half-plates'. The half-plates are divided by the building core containing toilets and lifts, etc. There is also a movement joint located in the core so the half-plates are individual structural entities. A considerable number of structural forms were considered for the building superstructures and the whole design team were involved in the selection to ensure that all factors were considered. The system adopted was a 400mm-deep waffle on a 1.5m-grid to suit the column grid of 7.5m and 9m-bays. The final constructed form was a system of in-situ spine and edge beams with precast rib beams and a profiled metal deck slab. The prime advantage of this system was the speed of construction. Only limited formwork was required for the spine and edge beams and the remainder of the floor slab construction was unpropped.

The roof structure was required to be as light as possible to minimise the visual impact on the space created. A system of lightweight tubular steel trusses was developed at close centres with purlins spanning between to carry the glazing rails. The size of the steelwork was minimised to such an extent that the glazing rails were noticeably larger. The glazing system was developed to have a minimal impact on transmitted light but to assist in the control of the temperature on warm days. This required the latest technology in glass coating systems to achieve the required performance without an undue cost penalty. The mass and spacing of the glass sheets were carefully calculated to exclude the maximum amount of aircraft noise and meet environmental requirements. The temperature of the street is tempered with exhaust air from the offices, some local heating around the café areas and local cooling on the bridges.

Opposite, clockwise from top: Early work on site; Comprehensive site plan; Main project model showing the relationship of the street and the six houses.

Above, from top: View of the Model; Overall site construction.

Museum of British History, St Barts

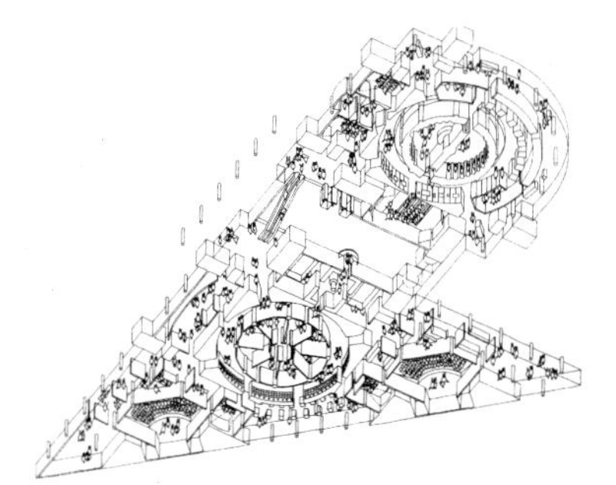

This is a most unusual museum based on five themed galleries covering British Landscape, British Invention, British Culture and the English language, Politics and Monarchy and the British People. From the outset the architects, display designers (Derek Walker Associates) and engineers (Buro Happold) knew that flexibility and integration were going to be the key elements in a narrative presentation where the display requirements would generate the building form. Therefore, the new museum building has been developed around the environmental and spatial needs of the five thematic galleries and the overview gallery – their interrelationship, preferred visitor circulation patterns and maximum accessibility to the site and building for the general public. The resulting optimum arrangement of four story of accommodation above a basement level places the building well below the suggested site lines to St Paul's.

Floor-to-floor heights within the galleries have been set at 6m. Opportunities exist at third-floor level to increase height in specific zones requiring greater volume. The two buildings containing the galleries are interlinked by a glazed atrium running their full height, within which are placed dynamic, temporary and permanent displays. The gallery floors are placed above a freely accessed ground floor in the form of an enclosed piazza which generates the main museum shop, main catering provision, temporary exhibition and display area, the main auditorium and the museum's information and ticketing facilities. The main museum building overlooks the central square of the St Bart's complex, and the eighteenth-century hospital buildings are integrated into the presentation and utilised for education, administration and corporate hospitality.

The proposed structural solution responds to a number of functional requirements. It comprises heavy superimposed loadings, adaptable mechanical and electrical servicing for flexible and varying use of space, adequate thermal mass to provide a moderate internal environment, high-quality soffits to central exhibition spaces, and a column grid and floor plan structure providing large clear span spaces without major constraints. The structural solution has also been chosen for speed of construction and its inherent fire resistance.

The southern entrance of the building envelope is characterised by glazed façades to interact visually with the external public space. Structural support to these elements must respond to this interaction while remaining visually unobtrusive. This has been achieved by utilising a curved form to generate a stable network of light rods capable of resisting any out-of-plane forces. Vertical loads are carried by the glazing which is suspended at floor level allowing the structure behind to remain as light as possible.

Museums are inherently high energy users. This in turn contributes to its environmental impact. The design addresses this problem. The main measures to facilitate energy saving rely on being able to tailor the environmental requirements of each individual space and to maintain this flexibility throughout the museum. In addition, the concrete soffits are exposed and will act as a thermal sponge stabilising temperatures in the space by absorbing excess heat. The whole complex includes not only the main museum but also conference, storage, preservation, restaurant and retail facilities. There is a benefit in generating a proportion of electricity on site and utilising waste heat to provide heating and, via absorption chillers, cooling for the building.

Opposite, top: Architectural model of the main museum building from the courtyard of St Barts (the two Georgian buildings will be used for administration and the museum's education department); Bottom: Approach to the museum from St Paul's showing the internal video wall.

Above, from top: Sketch of the CD Rom area in the Overview Gallery; Animated isometric of the gallery covering British Culture and the English Language and the gallery addressing Politics and the Monarchy; Model of the project showing the Gallery of British Landscape.

Al Faisaliah Complex, Riyadh

This project is a mixed-use development encompassing an office tower, five-star luxury hotel, apartments, retail mall and banquet hall. The scheme combines a careful balance of cost-effectiveness, flexibility and architectural interest in order to produce high quality buildings which will be modern, efficient in services, planning and operation, easily maintained and responsive to the Middle Eastern climate.

The building's layered façades will provide maximum control over the internal environment. The office tower rises to over 250m in height and is square in plan. Tapering to a point at the top in one giant arc it is designed around a compact central core with four main corner columns defining its unique silhouette. At stages up the building, observation decks are highlighted by giant K-braces which define the structure and tie the corner columns together. The building will be clad in natural silver, anodised aluminium panels with large, cantilevered sun-shading devices to minimise glare and to allow the use of non-reflective, energy-efficient glass. A vast banquet hall, designed to accommodate activities ranging from Islamic wedding ceremonies for up to 1,200 persons to conventions and trade shows, has been sunk below the landscaped plaza. An extraordinary degree of flexibility will be achieved by a unique long span arch system, allowing a column-free space of 57m x 81m.

On opposite sides of the plaza lie the hotel and apartment buildings, designed using solid indigenous materials – high-quality precast concrete, local limestone and wood – to create a multi-layered façade. The 40,000 square-metre retail mall flanking the central complex will have a 250m long central atrium punctuated with controlled natural lighting and a sophisticated projector/ lighting display at night.

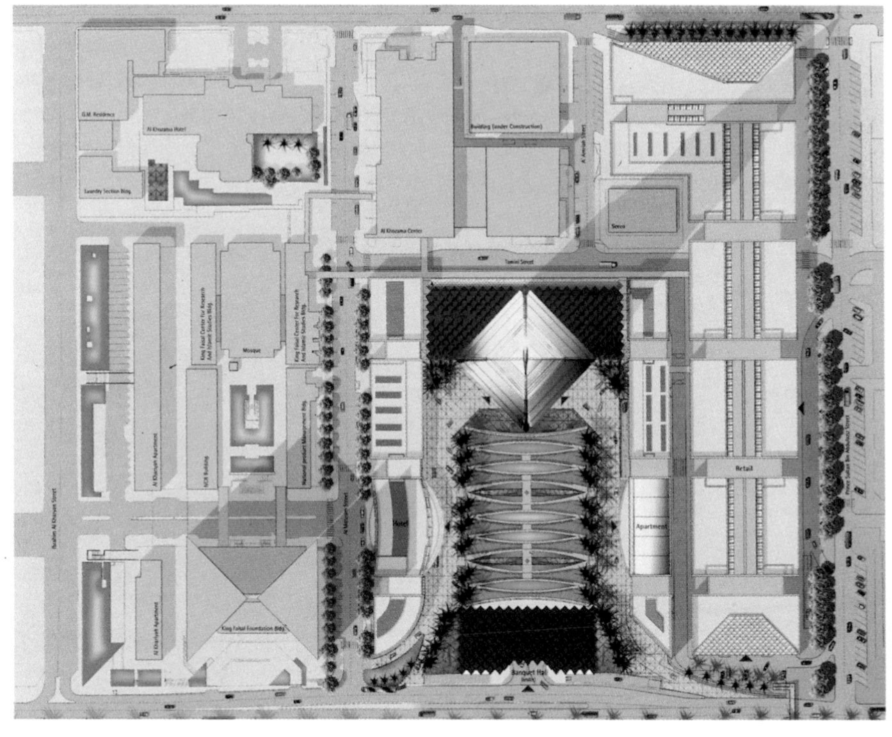

Opposite: Model of the Al Faisaliah complex, Riyadh.
Sir Norman Foster and Partners architects, Buro Happold engineers; Right: Interior model and site plan.

Epilogue

The English language, observed Ralph Waldo-Emerson, is 'the sea which received tributaries from every region in Heaven'. On a much smaller scale, the same might be said about 'the singular life' of Ted Happold. His deep personal convictions, his love of family, his unswerving commitment to his friends and colleagues, the width of his intellect, the global nature of his work, the quality of his engineering and his uncanny ability to create, communicate and inspire diverse teams to produce outstanding work – all contributed to the notion that Ted was a catalyst for an extended family that enjoyed his unique blend of excitement, devotion and good humour. Many of our great men are supremely unattractive outside their speciality: Ted's great strength was his 'chutzpah' which was all embracing. There was little or no division between work or play. He enjoyed life and transmitted that enjoyment to all those around him. To him, life was a great adventure and no conversation, trip, meal, design session or party was complete without the Happold factor – mischief, laughter, insight and generosity of spirit in equal measure.

This book attempts to capture the essence of the man. Michael Barclay, a close colleague at Arups, remembers Ted with great affection as a friend and mentor and, above all, an outstanding leader. He said 'I recognise him as a beacon of imagination, intellect and idealism in our rather dim industry. That he could also be devious, ruthless, and even unprincipled in pursuit of fame and fortune seems a small thing now beside his achievements and the warmth he gave to those close to him. He was an engineer who, like Arup, gave architects and engineers the support they needed to produce their best work and he picked some of his generation's best engineers to help realise their shared dreams ... Ted never rested on his laurels. Even during his address to the Institution of Stuctural Engineers, when he received the Gold Medal, he agreed with Duncan Michael's observation that the Gold Medal was not the end of one's career, but a mid-course school report covering a period to review where you had progressed, with the added responsibility to keep your medal clean".'

And, of course, for Ted that was never a problem. He retained to the end – like another old friend, Konrad Wachsmann – a delight and confidence in the young, a streak of individualism that fuelled his curiosity and endless search for new challenges. Like Wachsmann, he believed very much in the future, coupled with the belief that science and technology could help renew a construction industry that had become moribund, losing its skills in traditional methods of building and reluctant to spend resources in the development of complementary technological skills.

What is apparent is that the human world, past and future, is shaped by constant pressure from the imagination and ambition of individuals who have 'big ideas' – ideas that will prevail not by decree or even by persuasion, but because they capture the enthusiasm of people who will struggle against great difficulties to make them happen. In his book *Spaceships of the Mind*, Nigel Calder cites the disastrous consequences of boredom on the human psyche. To the visionary physicist, the late J D Bernal, daring was the essence of life. The astronomer Frank Drake commented 'There is no point keeping human beings healthy if there is nothing adventurous, nothing fun for them to do.' Ted's legacy is expressed in similar sentiments. He had an adventurous life, he was never bored or boring, he pursued the 'big idea' and left for his partners a binding fellowship that adopted his standards and ideals, and he generated a body of work that will not diminish in quality with his premature absence from the tiller.

The partners 1996 (Absent - Eddie Pugh and Terry Ealey);

Buro Happold personnel, January 1997: Rizal Abdulrahman Doloroso Adalim Phyllis Agbasi Gem Ahmet Mohammed Alauddin Roy Alimonsurin Rob Amphlett Bo Ascot Jo Austwick Babu Gurpal Bahra Bill Barraclough Nick Barwood David Beck Renet Bellow Neil Billett Nicole Breidohr Lee Bricker David Brocklehurst Peter Brooke Stephen Brown Sharon Bryant Jon Bull Tracy Cain John Calland Lito Capistrano Lisa Carbines Ken Carmichael Grant Carter Richard Carter Anish Chacko Nick Challand Richard Channing John Chanson Neil Clements Robert Clough Caroline Collier Andrew Comer **Mike Cook** David Cowles Maureen Crompton Neils Cross Paul Cullen Gary Cutler Mark Da Costa Gary Dagger Alan Dalton Alex Dando John Darch Andrea Dardin Nigel Davies Nils Den Hartog Peter de Boeck **Michael Dickson** Andy Dunford Joel Dunmore Teodoro Duro **Terry Ealey** Martin Eilers Helen Elias Andrew Fearnside Mark Fletcher Wendy Frampton Neil Francis Leon Furness Clive Fussell Denzil Gallagher Brian Gash Lucinda Gifford Colin Gill Colin Gimblett Vincent Grant **Mick Green** Nick Greenwood Richard Gregory Steve Gregson Vincent Grimley Simon Hammond Alan Harbinson Richard Harris Mike Harrison Janet Heaney Angela Heath Mathew Hele Linde Hillier Nigel Hiorons Tristram Hope Dave Hull Mathew Innes George Ittoop David James Phil James Carolyn Jones Heather Jones Mark Jones Eddie Jorden Martin Kealy **Padraic Kelly** Dennis Killick Dave Kingstone Emily Knight Mike Knott Harold Knowlson Krishna Kumar Roy Kurien Phil Lawrence Ian Leaper David Leversha Dominic Liddell **Ian Liddell** Carole Light Christine Lindley Paul Lipscombe Sui Ming Louie Mathew Lovell **Rod Macdonald** Steve Macey Ian Maddocks Tina Maggs Duncan Malins Tim Mander Roddy Manson Heather Marsden Emily McDonald **Tony McLaughlin** Jacqui McNicol Manoharan Moid Mansoor Abraham Mathew Kate Miller Mark Mitchell Morteza Mohammadi Andrew Mole John Morgan **John Morrison** Anna Morton **Peter Moseley** Andrew Murray Sasi Nair Peter Naylor Nick Nelson Mathew Norman Jon Norton Martin O'Brien Natasha Odlum Robert Okpala Omanakuttan Harry Orr Marilynn Osment Abdul Rahim D. Ostol Angus Palmer Ramesh Pandey Dawood Pandor Jeff Parkes Umesh Patel John Pearce Pam Penkman Birgit Pfortner Justin Phillips Kevin Phillips Eddie Picton Gloria Porter Richard Porter Sandy Porter Adrian Pottinger **Eddie Pugh** Lynnette Rangasammy Julia Ratcliffe Angela Reynolds Paul Richards Peter Roberts Leyla Rogers Paul Rogers Tammy Rogers Paul Romain Raini Rosnah-Ahmad Tanya Ross James Rowe Ather Saadi Manuel Salice Satheesh Craig Schwitter Peter Scott David Sergeant Mike Shaw Scott Shearer Julie Skailes Tom Skailes Sreekumar Allan Smith Carmel Smith Tony Soper Martin Spencer Samantha Spencer Neil Squibbs Sam Stacey Anne Stanislous Nicky Tanner Phil Tanner Stephen Tanno Lynne Taylor Mike Terriere Gavin Thompson Barbara Towers Iain Trent Glyn Trippick Tracy Tugwell Michael Vitzthum Simon Wainwright Tim Warren Chris Watson **Rodger Webster** Paul Westbury Jenny Wheelwright Hugh Wildy Ian Williams Tony Williamson Russel Winser Dave Wookey Romeo Yap Jian Kai Yi Gee Varghese Yohannan Jerry Young Saji Zacariah.

Selected Writings by Ted Happold

E Happold (1963) **New scheme to save Abu Simbel Temples**, *Architects' Journal* 20th March v.137 no.12, pp.610-612

E Happold & Poul Beckmann (1966) **Academic training for the building engineer,** *Proc. 2nd Conference on Teaching Engineering Design* Institution of Engineering Designers, 1316 April, Scarborough pp.60-66

E Happold & T Dannatt (1966) **Bootham School Assembly Hall,** *The Arup Journal* vol.1, no.1, March, pp.812

E Matchett & E Happold (1969) **The teaching and training of civil engineering design,** *Proc.Instn Civ.Engrs* vol.42, March, pp.451-454

E Happold (1970) **British Embassy, Rome,** *The Structural Engineer* vol.48, no.11, November, pp.417426; disc. vol.49, no.6, June 1971 pp.247-254

E Happold (1970) **British Embassy, Rome,** *The Arup Journal* vol.5, no.4, December, pp.1118

E Happold (1970) **The reconstruction of the Household Cavalry Barracks, London,** *Proc. Annual Congress, Public Works and Municipal Services* pp.226

E Happold (1971) **Hyde Park Cavalry Barracks,** *The Arup Journal* vol.6, no.3, September, pp.27

E Happold & P Rice (1973) **Centre Beaubourg,** *The Arup Journal* vol.8, no.2, June, pp.23

E Happold et al. (1973) **Long span cable roof structures** (discussion) *Proc.Instn Civ.Engrs* Part 1, vol.54, August pp.507-524

E Happold, M Barclay, J Martin & B Watt (1974) **Working in France,** *The Structural Engineer* vol.52, no.1, pp.316; disc. v.52 no.10, pp.381-388

E Happold & I Liddell (1974) **Blackheath Meeting House,** *The Arup Journal* vol.9, no.1, March, pp.21-23

E Happold [Structures 3] (1975) **Some roofs,** *Proc. 2nd Int. Conf. on Space Structures, University of Surrey* pp.469-485

E Happold (1975) **Timber lattice roof for the Mannheim Bundesgartenschau,** *The Structural Engineer* vol.53, no.3, March, pp.99135

E Happold and W I Liddell (1975) **Bundesgartenschau, Mannheim,** *The Arup Journal* vol.10, no.3, pp.11-18

E Happold (1975) **Riyadh Conference Centre and Hotel,** *The Structural Engineer* vol.53, no.12, pp.515-536

E Happold, M Courtney, J Pallaway & B Duck (1976) **To produce a building the relationships between contractor and professional consultants,** *Site Management Information Service* no.65, Spring, pp.15 (republished Institute of Building, 1980)

E Happold, I Liddell & M Dickson (1976) **Design towards convergence: a discussion,** *Architectural Design* vol.46, no.7, pp.430-435

E Happold (1977) **Beaubourg: architecture or engineering?** *Architectural Design* vol.47, no.2, pp.128-133

G Vincent & E Happold (1978) **How to integrate M&E services,** *Building Technology & Management* vol.16 no.2 February pp.911, 14

E Happold (1980) **Happold's Burocracy,** *RIBA Journal* vol.87, no.6, pp.46-48

E Happold & M Dickson (1980) **Airsupported structures review of the current state of the art contained in Airsupported structures,** *Institution of Structural Engineers*, pp.5-22

E Happold (1980) **Building under extreme conditions and development of appropriate construction technologies,** *Proceedings of 11th Annual Congress of IABSE,* Vienna pp.95-96

E Happold (1981) **Materials and components,** *Report of IABSE Symposium, London – The selection of structural form* pp.165-174

E Happold (1981) **The developing challenge for structural designers,** *The Structural Engineer* vol.59A, no.2, pp.5455; disc. vol.60A, no.7, pp.222-230

E Happold (1981) **Services included: the coordination of building services with the design of the building itself,** *Building* 30th October vol.241 no.7214 pp.24-25

E Happold & E Villefrance (1982) **Interaction between architect and structural engineer,** *IABSE Periodica* vol.2, May, pp.15-26

E Happold (1983) **Design and the profession,** *The Structural Engineer* vol.61A, no.10, pp.312-313

E Happold (1984) **The nature of engineering design: a conference for students,** *IABSE Periodica* vol.4, pp.153-156

E Happold, J F Woodward & J A Bergg (1984) **Can practising engineers teach and can teachers practise?** *Proc.Instn Civ.Engrs* Part 1, vol.76, August pp.804-806

E Happold et al. (1984) **Timber in building** *Building* 20th April vol.246 no.7339 pp.67-76

M Cook & E Happold (1984) **A survey of air supported structures** pp.710 in *The design of airsupported structures,* Institution of Structural Engineers, London

E Happold (1984) **The breadth and depth of structural design** pp.1622 in *The art and practice of structural design,* Institution of Structural Engineers, London

E Happold (1984) **Protectionism, liability and money,** *RIBA Transactions* vol.3 no.1 (5) pp.22-27

E Happold & S Macvicar (1984) **Educating builders,** *Building* 2nd November vol.247 no.7367 pp.30-31
E Happold (1985) **The conflict between classical and romantic,** *Deutsche Bauzeitung* vol.119 no.5, pp.10-15
A Faulkner, A Day & E Happold (1985) **Patterns of relationships in the building team,** *Building Technology & Management* vol.23 no.11 pp.28-30
E Happold (1986) **Presidential Address Can you hear me at the back?** *The Structural Engineer* vol.64A, no.12, December, pp.367-378
A Day, A Faulkner & E Happold (1986) **Computers and the Organisation of Design** *Proceedings of IABSE Workshop on Organising the Design Process, Zurich* IABSE Report No.53, pp.35-46
A Day, A Faulkner & E Happold (1986) **Communications and computers in the building industry** Construction Industry Computing Association, Cambridge
E Happold (1987) **The role of the professional: an engineers perspective,** *Design Studies* vol.7 no.3 July, pp133-138
E Happold (1987) **A personal perception of engineering,** pp.2439 in *Great Engineers* by Derek Walker, Academy Editions, London, 1987
E Happold (1987) **The nature of Engineering and Engineering in nature,** *Pidgeon Audio Visual* (tape slide pack)
E Happold, M Dickson & I Sutherland (1987) **A School for Woodland Industry,** *Proc.Int.Conf. on the Design & Construction of Nonconventional Structures Section V* Ed. B H V Topping, Civilcomp Press pp.107-114
E Happold (1987) **Building Industry Council new hope for the future,** *The Structural Engineer* vol.65A, no.10, October, pp.388
E Happold (1987) **Can we offer support?** (Introduction to conference 'Shelter and cities: building tomorrow's world') *The Structural Engineer* vol.65A, no.8, pp.281-283
E Happold (1987) **The Bossom Lecture: Some thoughts on architectural education,** *Public Health Engineer* vol.14, no.5
E Happold, T A Ealey, W I Liddell, J W E Pugh & R H Webster (1987) **The design and construction of the Diplomatic Club, Riyadh,** *The Structural Engineer* vol.65A, no.1, January, pp.15-26
E Happold (1987) **Education and the routes to chartered membership,** *The Structural Engineer* vol.65A, no.5, pp.161-179
E Happold (1987) **Presidential visit to Canada** (text of address marking centenary of Canadian Society of Civil Engineering) *The Structural Engineer* vol.65A, no.7, pp.264, 276
E Happold (1987) **Ten years overseas,** *Patterns no.1,* October, pp.23
E Happold (1988) **We ...,** *RSA Journal* vol.136 no.5380 March, pp.226-236
E Happold (1988) **The changing role of the consultant engineer in the UK,** *Patterns* no.2, April, pp.23
E Happold (1988) **Tensile building development,** *Proc. 1st Int. Oleg Kerensky Memorial Conf.* Session 3 pp.15
E Happold (1989) **Chariots of fire** *Patterns* no.5, May, pp.27
E Happold, J Morrison & T Ealey (1990) **Tsim Sha Tsui Cultural Centre. Inception, design and construction of the Hong Kong Cultural Centre,** *Proc.Instn Civ.Engrs* Part 1, October, pp.753763; disc. ICE Proceedings, Structs & Bldgs vol.94 Aug 1992 pp.351-357
E Happold, J Morrison & T Ealey (1990) **Tsim Sha Tsui Cultural Centre, Hong Kong the design concept,** *Patterns* no.9, October, pp.25
E Happold (1990) **Frei Otto: the force of nature,** *World Architecture* no.8, pp.36-44
E Happold & J Froud (1990) **A case study in appraisal and reuse,** *Patterns* no.7, December, pp.24
E Happold (1990) **Directions and opportunities,** *The Structural Engineer* (Special issue: 2nd Int'l Oleg Kerensky Memorial Conf.) May, pp.55-57
E Happold (1990) **Education for the single market,** *Building* 7th September vol.255 no.7663 pp.55-56
R Lutz, T McLaughlin, M Dickson & E Happold (1991) **The Commerzbank, Frankfurt a proposal for a green headquarters tower building,** *Patterns* no.10, December, pp.12-15
E Happold (1992) **Gold Medal Address: The nature of engineering** *The Structural Engineer* vol.70, no.10, October, pp.349-354
E Happold & W I Liddell (1992) **Tension Structures,** *Proceedings of 14th Annual Congress of IABSE,* New Delhi, pp.137-150
W A Allen, R G Courtney, E Happold & A Muir Wood (Eds.) (1992) **A global strategy for housing in the third millenium Spon,** London
E Happold (1993) **A journey in Saudi Arabia,** *Patterns* no.11, August, pp.27
E Happold (1995) **Chariots of fire. Tension structures a brief history,** *Architectural Design* vol.65, no.9/10, September/October, pp.30-35
E Happold (1996) **Conceptual design,** *Proc. of International Symposium: Conceptual design of structures* International Association for Shell and Spatial Structures / University of Stuttgart, August, pp.xix-xxiv

List of Projects Illustrated

Projects with Ove Arup & Partners

- Coventry Cathedral. 1955-1960. Architect: Sir Basil Spence. Page 36.
- Exeter University Science Building. 1960-1965. Architects: Sir Basil Spence, Bonnington and Collins. Page 37.
- Cavalry Barracks, Knightsbridge. 1960-1970. Architects : Sir Basil Spence Partnership. Page 36.
- British Embassy, Rome. 1961-1971. Architects: Sir Basil Spence Partnership. Page 39
- Bootham School Assembly Hall. 1962-1964. Architect: Trevor Dannatt & Partners. Page 46.
- Kennington Mixed Use Development. 1963-1965. Architect : Ted Hollamby, Borough of Lambeth. Page 43.
- Central Hill Housing. 1963-1973. Architect: Ted Hollamby, Borough of Lambeth. Page 40.
- Vauxhall Children's Lavatory. 1964-1965. Architect: Ted Hollamby, Borough of Lambeth. Page 41.
- West Norwood Library and Meeting Hall. 1964-1969. Architect: Ted Hollamby, Borough of Lambeth. Page 43.
- Riyadh Hotel and Conference Centre, Saudi Arabia. 1966-1973. Architects: Trevor Dannatt & Partners. Page 48.
- Crucible Theatre, Sheffield. 1967-1971. Architects: Renton Howard Wood Levin. Page 45.
- Mecca Conference Centre, Saudi Arabia. 1967-1973. Architects: Rolf Gutbrod with Frei Otto. Page 58.
- Kuwait Sports Centre. 1969. Architects: Kenzo Tange with Frei Otto. Page 65.
- Blackheath Meeting House. 1970-1972. Architects: Trevor Dannatt & Partners. Page 25.
- Warwick University Arts Centre. 1970-1974. Architects: Renton Howard Wood Levin. Theatre Consultant: John Bury. Page 45.
- Kensington Town Hall. 1970-1977. Architects: Sir Basil Spence, Bonnington and Collins. Page 38.
- Arctic City. 1971. Architects: Kenzo Tange with Frei Otto. Page 62.
- Beaubourg, Centre Pompidou, Paris. 1971-1975. Architects: Piano Rogers. Page 50.
- St Katherine's Dock, Tower Bridge re-development. 1972-1973. Architects: Renton Howard Wood. Page 44.
- Myatt's Field Boiler House. 1973-1974. Architect: Ted Hollamby, Borough of Lambeth. Page 40.
- Mannheim Bundesgartenschau. 1973-1975. Architect: Buro Mutschler & Partners with Frei Otto. Page 88.
- Kocommas Kings Office and Council of Ministers, Riyadh, Saudi Arabia. 1974-1977. Architects: Rolf Gutbrod, Buro Gutbrod with Frei Otto. Co-Engineers and Project Managers: Ove Arup & Partners. Co-Engineers : Buro Happold. Page 68.
- City Club, Milton Keynes. 1974-1978. Architects: Derek Walker Associates. Page 157.
- Dyce Tent for BP. 1975. Architects: Design Research Unit with Frei Otto. Page 64.
- Shadow in the Dessert, Muna Valley, Mecca, Saudi Arabia. 1975-1977. Architects: Rolf Gutbrod with Frei Otto. Page 75.
- Qatar University. 1975-1985. Architects: Dr Kamil El Khafrawi and Renton Howard Wood. Page 82.

Projects with Buro Happold

- British Genius Tent, Battersea. 1976. Architects: Pentagram. (Theo Crosby.) Page 110.
- Vail Ice Hockey Arena, Colorado. 1976-1977. Architects: Dale Sprankle and Bob Sprague with Ernie Kump. Page 101
- King Abdul Aziz University Sports Centre, Jeddah, Saudi Arabia. 1976-1979. Architects: Rolf Gutbrod, Buro Gutbrod with Frei Otto. Engineers: Buro Happold and Ove Arup & Partners. Page 64.
- Cabrio Folding Stand Cover. 1977. Architect: Frei Otto. Page 67.
- Pink Floyd Umbrellas, American Tour. 1978. Co-Designer: Frei Otto. Page 67.
- Munich Aviary. 1978-1982. Architects: Jorg Gribl with Frei Otto. Page 130.
- Museum of Archaeology, Jordan. 1979. Architect: Michael Brawne Associates with Arab Tech, Jordan. Page 152.
- Sculpture Park, Milton Keynes. 1979. Architects: Derek Walker Associates. Page 153.
- Al Marzook Mosque and Centre for Islamic Medicine, Kuwait. 1979-1987. Architect: Isaac Zekaria. Page 83.
- San Diego Aviary. 1980-1981. Architects and Co-Engineers: Buss Silvers with Jorg Gribl. Page 34.
- The Sainsbury Extension, Worcester College, Cambridge. 1980-1982. Architects: McCormac, Jamieson and Prichard. Page 116.
- 58ºN. 1981. Architects: Arni Fullerton, Dennis Wilkinson with Frei Otto. Page 150.
- Baltimore Music Pavilion. 1981. Designers and Engineers: FTL and Buro Happold. Co-Engineers: Bob Silman Associates. Page 72.
- British Embassy, Riyadh. 1981-1986. Architects: Trevor Dannatt & Partners. Page 49.
- Tsim Sha Tsui Cultural Centre, Hong Kong. 1981-1989. Architects: Public Works Department, Hong Kong (Jose Lei). Hong Kong Engineers: Ho Happold. Page 106.
- Vauxhall Cross (Green Giant Project), London. 1982. Architects: Sebire Allsopp. Page 148.
- Canopy for the Washington Symphony Orchestra. 1982. Designers and Engineers: FTL and Buro Happold. Page 61.
- British Telecom Headquarters, Milton Keynes. 1983. Architects: Ahrends Burton Koralek. Page 129.
- Fleet Velmead Infants School, Hampshire. 1984-1986. Architects: Michael Hopkins Partnership. Page 118.
- Quba Mosque, Medina, Saudi Arabia. 1984-1986. Architect: Abdul Wahib El Wakil. Page 81.
- Kasr Al Hokm, Riyadh, Saudi Arabia. 1984-1992. Architects: Shubeilat Badran Infrastructure Saud Consult for the Arriyadh Development Authority. Page 76.
- Grand Mosque, Riyadh. 1984-1992. Architects: Shubeilat Badran Associates. Page 80.
- Leeds Playhouse. 1985. Architects: Derek Walker Associates. Theatre Consultant: John Bury. Page 124.
- Energy Pavilion, WonderWorld, Corby. 1985-1986. Architects: Derek Walker Associates. Page 157.
- Ocean Park Aviary, Hong Kong. 1985-1986. Architects: Leigh & Orange. Co-Engineers: Ho Happold. Page 104.
- State Mosque, Sarawak. 1985-1989. Architects: Sami Mousawi and Peruniding Utama. Page 85.
- Hooke Park, Dorset, Phase 1. 1985-1991. Architects: Ahrends Burton Koralek (Richard Burton) with Frei Otto. Page 130.
- Airfish. 1986. Frei Otto and Buro Happold. Page 62.
- Solid State Logic, Oxford. 1986-1988. Architects: Michael Hopkins Partnership. Page 114.
- Kowloon Park, Hong Kong. 1987. Architects: Derek Walker Associates. Hong Kong Engineers: Ho Happold. Page 106.
- Airship 1. 1987-1988. Frei Otto and Buro Happold. Page 73.
- High Wycombe Arts Centre. 1987-1988. Architects: Derek Walker Associates. Theatre Consultant: John Bury. Page 127.
- Imagination Building, Stafford House, Store Street, London. 1987-1989. Architect: Ron Herron. Page 138.
- Bishopstoke School, Hampshire. 1987-1989. Architects: Sir Colin Stansfield-Smith, Hampshire County Council. Page 120.
- Pearls of Kuwait Competition. 1987-1990. Alternate Architects: Jerde Partnership, Los Angeles and Philip Cox, Sydney, Australia. Page 154.
- The Katherine Hamnett Shop, Brompton Road, London. 1988. Architects: Foster Associates. Page 144.
- Hong Kong Technical University. 1988-1989. Architects: Derek Walker Associates. Page 105.
- Hazelwood Primary School, West Totton, Hampshire. 1988-1990. Architects: Sir Colin Stansfield-Smith, Hampshire County Council. Page 121.
- Bedfont Lakes, IBM Building. 1989-1993. Master Planners and Architects: Michael Hopkins Partnership. Page 115.
- Bedfont Lakes Northern Offices. 1989-1993. Architects: Edward Cullinan & Partners. Page 136.
- Globe Theatre, Southwark, London. 1989-1996. Architects: Pentagram. (Theo Crosby.) Page 111.
- Wellington Gate Visitors Centre, Kew. 1990-1991. Architects: Trevor Dannatt & Partners. Page 49.
- Offices and Extension of Research Laboratories, Schlumberger, Cambridge. 1990-1992. Architects: Michael Hopkins Partnership. Page 112.
- Diplomatic Club, Riyadh. 1991-1992. Architects: Omrania (Nabil Fanous and Basem Shihabi) with Frei Otto. Page 122.
- Hyperion Airship, Disneyland, Paris. 1991-1992. Engineers: FTL and Buro Happold. Page 73.
- Covering to Town Square, Basildon Town Centre. 1991-1993 Architects: Michael Hopkins Partnership. Page 60.
- Transrapide Magnetbahn Project, Germany. 1992. Co-Designer: Frei Otto. Page 70.
- Boston Harbour Lights. 1992-1994. Engineers: FTL and Buro Happold. Co-Engineers: Dalland Associates. Page 73.
- Headquarters for RWE, Essen. 1992-1994. Architects: IOP (Christoph Ingenhoven). Page 146.
- Queens Building, Emmanuel College, Cambridge. 1992-1995. Architects: Michael Hopkins Partnership. Page 114.
- Techniquest, Cardiff. 1992-1995. Architects: Ahrends Burton Koralek (Paul Koralek). Page 128.
- Pneumatic Maintenance Building for Helicopters, US Army. 1993-1994. Engineers: FTL and Buro Happold. Page 73.
- Eastleigh Tennis Centre. 1993-1994. Architects: Euan Burland Assoicates. Page 158.
- International Manufacturing Centre Bridge, Warwick University. 1993-1995. Architects: Edward Cullinan & Partners. Page 136.
- Prophets Holy Mosque Umbrellas, Medina. 1993-1995. Architects: Bodo Rasch with Sonder Konstructien Leichtbau. Page 66.
- Royal Armouries Museum, Leeds. 1993-1996. Architects: Derek Walker Associates. Page 140.
- Prospect Park Combined Business Centre for British Airways, Heathrow. 1993 onwards. Architect: Nils Torp with Renton Howard Wood Levin. Construction Managers: Mace. Page 166.
- Westminster Lodge, Hooke Park, Dorset. 1994-1996. Architects: Edward Cullinan & Partners. Page 136.
- Al Faisaliah Complex, Riyadh for the King Faisal Foundation. 1994 onwards. Architects: Sir Norman Foster & Partners. Page 170.
- St Peter's Church Renovation, Bristol. 1995 onwards. Architects: Ahrends Burton Koralek (Richard Burton) with Frei Otto. Page 129.
- Glasgow Millennium Wind Tower. 1995 onwards. Architect: Richard Horden. Page 156.
- Lowry Centre, Salford. 1995 onwards. Architects: Michael Wilford & Partners. Page 164.
- The Great Court, British Museum. 1995 onwards. Architects: Sir Norman Foster & Partners. Page 160.
- AT & T Building, Atlanta, 1996 Olympics. Designers and Engineers: FTL and Buro Happold. Page 158.
- Dome for the Millennium Experience. 1996 onwards. Architects : Richard Rogers Partnership. Page 162.
- Museum of British History. 1996 onwards. Architects : Derek Walker Associates. Page 166.